# LINEAR VECTOR SPACES AND CARTESIAN TENSORS

JAMES K. KNOWLES
*California Institute of Technology*

New York Oxford
**Oxford University Press**
1998

Oxford University Press
Oxford New York
Athens Auckland Bangkok Bogotá Bombay Buenos Aires
Calcutta Cape Town Dar es Salaam Delhi Florence Hong Kong
Istanbul Karachi Kuala Lumpur Madras Madrid Melbourne
Mexico City Nairobi Paris Singapore Taipei Tokyo Toronto

and associated companies in
Berlin Ibadan



Published by Oxford University Press, Inc.,
198 Madison Avenue, New York, New York, 10016
http://www.oup-usa.org
1-800-334-4249



**Library of Congress Cataloging-in-Publication Data**

Knowles, James K. (James Kenyon), 1931–
Linear vector spaces and Cartesian tensors / James K. Knowles.
p. cm.
Includes bibliographical references and index.
ISBN 0-19-511254-7 (cloth)
1. Vector spaces 2. Calculus of tensors. I. Title.
QA186.K657 1997
512′.52—dc21 97-8936
CIP

Printing (last digit): 9 8 7 6 5 4 3 2 1
Printed in the United States of America
on acid-free paper

*To Jackie, John, Jeff, and Jamey*

# TABLE OF CONTENTS


*Preface* *vi*

**1. Linear Vector Spaces** **1**
References and Problems *14*

**2. Linear Transformations** **18**
References and Problems *36*

**3. Finite-Dimensional Euclidean Spaces and Cartesian Tensors** **42**
References and Problems *59*

**4. 4-Tensors** **67**
References and Problems *74*

**5. Applications** **78**
References and Problems *94*

**Appendix 1. Assumed Background** **99**

**Appendix 2. Solutions for Selected Problems** **102**

*Index* *118*

# PREFACE

This book represents my approach to some applicable material connected with linear vector spaces that has been part of a course offered in the Division of Engineering and Applied Science at Caltech for many years. The course—called AM125 in the local code—covers a full academic year, which is divided into three quarters, with this material offered in the first quarter. AM125 is usually taught by different instructors in different years, and the material related to vector spaces looks quite different from one instructor to another.

The typical consumer in AM125 is a first-year graduate student, usually from a field in engineering or applied science, but sometimes from geology and occasionally from chemistry or physics. Students in many of the doctoral programs within the Division of Engineering and Applied Science are *required* by their elders to jump through this particular mathematical hoop. Although the backgrounds of the students vary widely, almost all have had something called "linear algebra" in their undergraduate programs, in addition to basic calculus, differential equations, and often complex variables. All of the takers have contemporaneous sustained exposure to serious science or engineering. As a result, most are quite well equipped to appreciate the potential utility of the material presented in AM125.

Usually, a sizable majority of the students who take AM125 are or will be pursuing doctoral research in fields that make heavy use of some aspect of mechanics, often the continuum mechanics of solids or fluids. The slant that I choose to put on the material, especially that covered in the first quarter, is intended to make the subject particularly helpful to this group, though not—I hope—to the detriment of others.

It has been my experience that the weapons brought by the students to AM125 from their earlier experience in linear algebra are for the most part limited to those that help in doing calculations with matrices, especially big ones. Every student can calculate with astonishing accuracy something he or she will call an eigenvalue, but frequently the understanding of what an eigenvalue actually *means* is dim or absent altogether. The geometric content of the theory of vector spaces is often missing from the student's knowledge, and notions of invariance, so important in continuum mechanics, for example, evidently have been given little attention. Most students do not appreciate the fundamental distinctions between vector spaces with real scalars and those whose scalars are complex.

It is my hope that this particular version of the theory and application of finite-dimensional vector spaces will provide remedies for the specific illnesses cited above, while also offering an interesting and useful look at a subject of great beauty and utility. Little is done here with infinite-dimensional spaces, though I have touched upon them from time to time. The basic material on finite-dimensional linear spaces is covered in the first three chapters. Chapter 4 introduces the subject of Cartesian tensors of rank four, partly because of their importance in continuum mechanics, but also as an application of the ideas to the construction of another useful mathematical device. Some applications that lie closer to real physical issues are covered in the final chapter. There are problems of varying degrees of difficulty at the end of each chapter. Solutions to most of these problems are sketched in Appendix 2, though solutions are not given for the very easiest problems, nor for those in which the student is given detailed instructions in the statement of the problem itself. Appendix 1 contains a brief list of some results from elementary algebra that are taken for granted throughout the book. A short list of relevant references is included at the end of each chapter.

I teach slowly. I have never covered literally *all* of the material presented here in one academic quarter, but I have no doubt that others could readily do so. It would be easy to do—even for me—in a semester.

A word about undergraduates: Although there are usually few undergraduates in AM125 at Caltech, this is by no means because they are unprepared for the mathematical material and its applications. Indeed, this book could serve as the text for an advanced undergraduate course, especially for students with some exposure to courses in science or engineering.

There are, of course, many fine books on the subject treated here. I must especially acknowledge two that I admire greatly and to which I have repeatedly turned for illumination for many years: One is *Lectures on Linear Algebra*, by I.M. Gel'fand, the other *Finite Dimensional Vector Spaces*, by P.R. Halmos. I owe both of these unique resources a great debt of gratitude.

A word of thanks to the many years worth of Caltech students who have taken AM125 either as conscripts or as volunteers: Their efforts, comments and criticisms have led to many revisions of the presentation and the problems. I hope at least some of them have had as much fun confronting the material as I have had in delivering it!

It is pleasant to acknowledge the countless conversations I have had over the years with colleagues at Caltech and elsewhere concerning the subject addressed here and matters closely related to it. I wish particularly to thank Jim Beck, Tom Caughey, and Steve Wiggins of Caltech, Rohan Abeyaratne of M.I.T., Mort Gurtin of Carnegie-Mellon University, Niall Horgan of the

University of Virginia, Phoebus Rosakis of Cornell University, and Dick Shield of Caltech and the University of Illinois for many interactions that contributed enormously to my understanding and appreciation of the issues. Finally, I must express my very special gratitude to the late Eli Sternberg of Caltech for years of exchanges concerning this subject and many others. Although Eli, alas, never taught AM125, his interest in linear spaces was intense, and he was a master of their applications in solid mechanics. I profited greatly from our joint ruminations over twenty-five years.

I am also pleased to acknowledge the efficient help and welcome encouragement that I have received from Mr. Bill Zobrist, who is my editor at Oxford University Press.

C H A P T E R 1

# LINEAR VECTOR SPACES

When we are young, we learn that vectors are arrows. We are told how to add arrows by the parallelogram law, how to multiply arrows by numbers, including negative ones, and how to measure the length of an arrow or the angle between two of them. We are shown how to rotate arrows, stretch them, and reflect them. These operations have to do with the arithmetic of arrows, the metric properties of arrows, and the transformation of arrows into other arrows according to some preset rule, such as rotation through an angle of 45° about a given axis. At this later stage of life, we might observe that the geometric definitions of these operations and the attendant calculations do not involve the use of "coordinate axes" along which to "resolve" the arrows into their "components." The significance of this observation will unfold as we go along.

The collection $\mathsf{A}_2$ of all arrows in a plane with their tails at a fixed point (the origin), together with the real numbers we need to multiply arrows by, represents the prototype of the useful abstract notion of a linear vector space. We can also deal, of course, with the set $\mathsf{A}_3$ of all arrows in ordinary *three*-dimensional space rather than the set of arrows confined to a plane. In any event, it is linear vector spaces that we shall explore in this book. Along the way, we shall encounter some applications of the theory, mainly to that most venerable of physical disciplines, *mechanics*.

A *linear vector space* $\mathsf{R}$ is a collection of things called *vectors* (denoted by bold face letters like $\mathbf{x}$, $\mathbf{y}$, $\mathbf{u}$, $\mathbf{v}$, ...), together with some things called scalars (usually designated by symbols like $\alpha$, $\beta$, a, b, ...), that jointly conform to certain rules that have been abstracted from the story of arrows. These rules can be conveniently grouped in two sets:

I. Given any two vectors $\mathbf{x}$ and $\mathbf{y}$ in $\mathsf{R}$, there is a third vector $\mathbf{z}$ in $\mathsf{R}$, called the sum of $\mathbf{x}$ and $\mathbf{y}$ and written $\mathbf{z} = \mathbf{x} + \mathbf{y}$, that obeys the following rules:

(a) addition is commutative ($\mathbf{x} + \mathbf{y} = \mathbf{y} + \mathbf{x}$) and associative ($(\mathbf{x} + \mathbf{y}) + \mathbf{w} = \mathbf{x} + (\mathbf{y} + \mathbf{w})$), etc.;

(b) there is a unique vector $\mathbf{o}$ in $\mathsf{R}$ (called the *null* vector) such that $\mathbf{x} + \mathbf{o} = \mathbf{x}$;

(c) for every vector $\mathbf{x}$ in $\mathsf{R}$, there is a unique vector called $-\mathbf{x}$ such that $\mathbf{x} + (-\mathbf{x}) = \mathbf{o}$.

II. Given any vector $\mathbf{x}$ and any scalar $\alpha$, there is a vector $\mathbf{z}$ called the product of $\mathbf{x}$ with $\alpha$ and written $\mathbf{z} = \alpha\mathbf{x}$, that obeys the following rules:

(a) multiplication by a scalar is distributive in the sense that $\alpha(\mathbf{x} + \mathbf{y}) = \alpha\mathbf{x} + \alpha\mathbf{y}$ and $(\alpha + \beta)\mathbf{x} = \alpha\mathbf{x} + \beta\mathbf{x}$;

(b) there are scalars called 0 and 1 such that $0\mathbf{x} = \mathbf{o}$, $1\mathbf{x} = \mathbf{x}$.

Although more general choices are possible, the objects referred to as *scalars* above are usually either real numbers or complex numbers. Our interest will ultimately lie in *real* vector spaces, i.e., spaces with real numbers as scalars, but we will also need *complex* vector spaces (spaces whose scalars are complex numbers) from time to time, especially for purposes of contrast.

The arrows in $\mathsf{A}_2$ or $\mathsf{A}_3$ comprise a real vector space; if $\mathbf{x}$ is an arrow, $3\mathbf{x}$ is the arrow three times as long as $\mathbf{x}$ whose direction is that of $\mathbf{x}$, and $(-1)\mathbf{x} \equiv -\mathbf{x}$ is the arrow directed exactly opposite to $\mathbf{x}$ whose length coincides with that of $\mathbf{x}$. You may find it helpful to verify that the spaces of arrows $\mathsf{A}_2$ and $\mathsf{A}_3$ conform to the rules for a linear vector space laid out above.

Here are some other examples of vector spaces.

***Example 1.1.*** *The space of columns of real numbers.* A much-loved vector space is the collection $\mathsf{R} = \mathsf{R}_\mathrm{n}$ of all columns $\mathbf{x}$ of n real numbers, where n is a positive integer; the scalars are real numbers. Thus a typical element of $\mathsf{R}_\mathrm{n}$ is of the form

$$\mathbf{x} = \begin{pmatrix} x_1 \\ x_2 \\ \cdot \\ \cdot \\ \cdot \\ x_n \end{pmatrix}, \tag{1.1}$$

where $x_1, x_2, \ldots, x_n$ are real numbers. If $\mathbf{x}$ and $\mathbf{y}$ are two vectors in $\mathsf{R}_\mathrm{n}$, their sum is defined to be the column whose entries are the sums of corresponding entries in $\mathbf{x}$ and $\mathbf{y}$ (i.e., $x_1 + y_1$, $x_2 + y_2$, etc.). Similarly, $\alpha\mathbf{x}$ is the column with entries $\alpha x_1, \alpha x_2, \ldots, \alpha x_n$. The null vector $\mathbf{o}$ is the column with every entry zero, and $-\mathbf{x}$ is the column in which the generic entry is $-x_k$. Under these operations, the collection $R_n$ is readily shown to obey the rules for a vector space set out above. When we use $R_n$ in the future, we shall refer to the operations of addition of vectors and multiplication of a vector by a scalar as specified here as the "natural" operations on columns.

Because Descartes explained in the seventeenth century that a plane can be usefully described in terms of the "coordinates" of each of its points—say $x_1$, $x_2$—with respect to a set of rectilinear axes, the special case $R_2$ of the vector space in Example 1.1 is particularly valuable. So of course is $R_3$, which describes the space we live in through the coordinates $x_1$, $x_2$, $x_3$.

***Example 1.2.*** *The space of columns of complex numbers.* An equally good example of a vector space is the set $C_n$ of all columns of n *complex* numbers $z_1$, $z_2$, ..., $z_n$, with the complex numbers serving as scalars. The definitions of the operations of addition and multiplication by a scalar are analogous to those used in the invention of $R_n$ in Example 1.1.

***Example 1.3.*** *The space of continuous real-valued functions.* Let C be the collection of all real-valued functions that are defined and continuous on the closed interval $[0, \pi]$. If f and g are two such functions, we say their sum $h = f + g$ is the real valued function defined by $h(t) = f(t) + g(t)$ for every t such that $0 \leq t \leq \pi$, and if $\alpha$ is a real number, $h = \alpha f$ is defined by $h(t) = \alpha f(t)$, $t \epsilon [0, \pi]$. C is a vector space under these "natural" operations.

***Example 1.4.*** *Solutions of a linear ordinary differential equation.* Consider the collection of all twice continuously differentiable, real-valued functions $\varphi$ on the interval $[0, \pi]$ that satisfy the linear ordinary differential equation

$$\varphi'' + \varphi = 0 \quad \text{on } [0, \pi]. \tag{1.2}$$

Let the scalars be real numbers, and define the operations of addition and multiplication by a scalar exactly as in the preceding example. Because the sum of two twice continuously differentiable functions is also twice continuously differentiable, because the sum of two solutions of (1.2) is also a solution, and because any real multiple of a solution is a solution, this collection is also a linear vector space. Observe that every vector in this vector space is also in the space C of the preceding example. Problem 1.1 explores how things would change if the zero on the right side of (1.2) were replaced by one.

***Example 1.5.*** *A space of polynomials.* If N is a non-negative integer and $n = N + 1$, denote by $P_n$ the set containing every polynomial with real coefficients whose degree is not greater than N. If $\mathbf{p}$ and $\mathbf{q}$ are two polynomials in $P_n$, define $\mathbf{p} + \mathbf{q}$ and $\alpha\mathbf{p}$, where $\alpha$ is real, in the natural way. $P_n$ then becomes a real vector space. Note that $P_n$ is contained in the space C of Example 1.3.

The first notions of importance in the study of vector spaces are the complementary ones of linear dependence and linear independence. Let R be a linear vector space, and let $\mathbf{x}_1, \mathbf{x}_2, ..., \mathbf{x}_k$ be a set of k vectors in R, where k is a positive integer. This set is said to be *linearly dependent* if there are k scalars $\alpha_1, \alpha_2, ..., \alpha_k$, *not all zero*, such that

$$\alpha_1\mathbf{x}_1 + \alpha_2\,\mathbf{x}_2 + ... + \alpha_k\,\mathbf{x}_k = \mathbf{o}. \tag{1.3}$$

If the only set of scalars for which (1.3) holds is the set $\alpha_1 = \alpha_2 = ... = \alpha_k = 0$, the set of k vectors $\mathbf{x}_1, ..., \mathbf{x}_k$ is said to be *linearly independent*. Note that linear dependence and independence are properties of *sets* of vectors, not of individual vectors themselves. The set $\{\mathbf{x}\}$ containing the single vector $\mathbf{x}$ is a linearly independent set unless $\mathbf{x} = \mathbf{o}$, in which case it is linearly dependent. Any set of k vectors that includes $\mathbf{o}$ is a linearly dependent set; see Problem 1.3.

In the vector space $A_2$ of arrows in a plane, any set of two arrows that are parallel (or anti-parallel) is a linearly dependent set. Any set of two arrows that are both of non-zero length and neither parallel nor anti-parallel is linearly independent. Any set of three or more arrows is linearly dependent; see Problem 1.4.

In the vector space of Example 1.4 above, $\varphi_1(t) = \sin t$ and $\varphi_2(t) = \cos t$ are vectors in the space, and together they clearly constitute a linearly independent set. The theory of linear ordinary differential equations tells us that *every* solution of (1.2) is expressible as a linear combination of these two special solutions. This means that any set of *three* elements of the vector space of solutions of (1.2) that contains both $\varphi_1$ and $\varphi_2$ is a linearly *dependent* set. Indeed, as discussed in Problem 1.5, the theory of differential equations tells us more than this: *any* set of three solutions of (1.2) is a linearly dependent set.

The vector space of Example 1.3 is richer. Let k be *any* positive integer. For every positive integer k, the set of continuous functions $\varphi_1(t) = 1, \varphi_2(t) = t, \varphi_3(t) = t^2, ..., \varphi_{k+1}(t) = t^k$ is a linearly independent set, as will be shown in Problem 1.6. Thus in this vector space, linearly independent sets can be as big as one likes, in the sense that such a set can contain an arbitrarily large number of vectors.

If a vector space R contains a linearly independent set of n vectors but contains *no* linearly independent set of n + 1 vectors, where n is a positive integer, then R is said to have dimension n. The set $\{\mathbf{o}\}$ consisting of the null vector alone is trivially a vector space; it fails to contain a linearly independent set of n vectors for *any* positive integer n, and is therefore said to have dimension zero. In contrast, there are vector spaces that contain linearly independent sets of n vectors for *every* positive integer n; such spaces are said to be infinite dimensional. For the most part, we shall be concerned with finite-dimensional spaces.

Let R be a vector space of finite dimension $n \geq 1$. A linearly independent set of n vectors is called a *basis* for R. Let $\mathbf{e}_1, \mathbf{e}_2, \ldots, \mathbf{e}_n$ be a basis for R, and let $\mathbf{x}$ be an arbitrary vector in R. Then, by the definition of dimension, the set $\{\mathbf{x}, \mathbf{e}_1, \mathbf{e}_2, \ldots, \mathbf{e}_n\}$ is linearly dependent, so there are n + 1 scalars $\alpha_0, \alpha_1, \ldots, \alpha_n$, not all zero, such that $\alpha_0\mathbf{x} + \alpha_1\mathbf{e}_1 + \alpha_2\,\mathbf{e}_2 + \ldots + \alpha_n\,\mathbf{e}_n = \mathbf{o}$. Clearly $\alpha_0$ cannot vanish, or the **e**s would comprise a linearly dependent set. For k = 1, ..., n, put $\xi_k = -\alpha_k/\alpha_0$. Then

$$\mathbf{x} = \xi_1\,\mathbf{e}_1 + \xi_2\,\mathbf{e}_2 + \ldots + \xi_n\,\mathbf{e}_n. \tag{1.4}$$

This shows that, for every $\mathbf{x}$ in R, there are scalars $\xi_1, \xi_2, \ldots, \xi_n$ such that $\mathbf{x}$ may be represented in terms of the basis $\mathbf{e}_1, \mathbf{e}_2, \ldots, \mathbf{e}_n$ by (1.4). The linear independence of the set of basis vectors assures that, for the given $\mathbf{x}$, the scalars $\xi_i$ in (1.4) are unique, as will be shown in Problem 1.7. They are called the *components* of $\mathbf{x}$ *in the basis* $\mathbf{e}_1, \mathbf{e}_2, \ldots, \mathbf{e}_n$, or *in the basis* $\mathbf{e}$, for brevity. The components of a vector are both useful and dangerous. Their utility stems from the fact that, as we shall see often enough, using components is frequently helpful in making calculations. The potential danger in appealing to components arises because they depend not only on the vector being represented, but also on the choice of basis. Thus when a result is established by a calculation involving components, one may question whether the result is basis-dependent. Because basis-*in*dependent results are crucial in many physical applications of our subject, one must keep this issue in mind as we proceed.

One further depressing note: the discussion in the preceding paragraph fails to explain how to *find* the components of a given vector. We shall address this question later.

Back to the space $A_2$ of arrows in a plane. Since according to Problem 1.4, any three arrows in $A_2$ are linearly dependent, the dimension of $A_2$ cannot exceed 2. But if $\mathbf{e}_1$ and $\mathbf{e}_2$ are two perpendicular arrows of non-zero length, neither is a scalar multiple of the other, so together they comprise a linearly independent set, and—since there cannot be three arrows with this property—they form a basis. Suppose in addition that $\mathbf{e}_1$ and $\mathbf{e}_2$ are perpendicular arrows, each with unit length, and let $\mathbf{x}$ be any arrow in $A_2$. If $\theta$ is the angle between $\mathbf{x}$ and $\mathbf{e}_1$ and $|\mathbf{x}|$ is the length of the arrow $\mathbf{x}$, some trigonometry shows that $\mathbf{x} = |\mathbf{x}| \cos\theta\, \mathbf{e}_1 + |\mathbf{x}| \sin\theta\, \mathbf{e}_2$, so that $\xi_1 = |\mathbf{x}| \cos\theta$ and $\xi_2 = |\mathbf{x}| \sin\theta$ are the components of the arrow $\mathbf{x}$ in this particular basis.

As a second illustration, consider the space $\mathsf{R}_2$ of columns of two real numbers. It is again easy to show that any three vectors in $\mathsf{R}_2$ are linearly dependent. Moreover, if

$$\mathbf{x} = \begin{pmatrix} x_1 \\ x_2 \end{pmatrix} \tag{1.5}$$

is any vector in $\mathsf{R}_2$, and

$$\mathbf{e}_1 = \begin{pmatrix} 1 \\ 0 \end{pmatrix}, \quad \mathbf{e}_2 = \begin{pmatrix} 0 \\ 1 \end{pmatrix}, \tag{1.5a}$$

clearly $\mathbf{e}_1$ and $\mathbf{e}_2$ form a linearly independent set and therefore a basis (often called the *natural* basis for $\mathsf{R}_2$). Furthermore, $\mathbf{x} = x_1\, \mathbf{e}_1 + x_2\, \mathbf{e}_2$, so that $\xi_1 = x_1$ and $\xi_2 = x_2$ are the components of $\mathbf{x}$ in this basis. This discussion has an easy generalization to $\mathsf{R}_n$, for any positive integer n.

Our earlier remarks lead us to conclude that the dimension of the vector space of solutions of the differential equation (1.2) in Example 1.4 is 2, and that cos t and sin t form a basis.

For the vector space $\mathsf{P}_n$ of Example 1.5, the special polynomials 1, t, $t^2$, ..., $t^N$ are linearly independent and form a basis; the dimension of $\mathsf{P}_n$ is $n = N + 1$. Problem 1.8 addresses the question of how to find the components in this basis of a polynomial $\mathbf{p}$ in $\mathsf{P}_n$.

Example 1.3 provides an example of an infinite-dimensional vector space; since the definition of basis given above applies only to finite-dimensional spaces, we cannot speak of this notion for the space of Example 1.3 without extending the idea in some suitable way. This task presents some troublesome technical problems that would take us well off our intended path, so we avoid this issue and suggest that the reader who is especially interested in infinite-dimensional spaces should consult the references by Halmos [1.4], Kolmogorov and Fomin [1.5], Debnath and Mikusinski [1.1], and Naylor and Sell [1.6] listed at the end of this chapter.

A question that will arise from time to time in what follows concerns the relationship between the components of a vector in two bases. Let $\mathbf{e}_1$, ..., $\mathbf{e}_n$ and $\mathbf{f}_1$, ..., $\mathbf{f}_n$ be two bases for a vector space $\mathsf{R}$ of finite dimension n, and let $\mathbf{x}$ be a given vector in $\mathsf{R}$. Then we may write

$$\mathbf{x} = \sum_{k=1}^{n} \xi_k\, \mathbf{e}_k = \sum_{k=1}^{n} \eta_k\, \mathbf{f}_k, \tag{1.6}$$

in terms of the two sets of components $\xi_k$ and $\eta_k$. But since the vectors $\mathbf{f}_k$

form a basis, each of the vectors $\mathbf{e}_k$ may be expressed as follows:

$$\mathbf{e}_k = \sum_{j=1}^{n} \rho_{jk}\, \mathbf{f}_j \tag{1.7}$$

for some scalars $\rho_{jk}$; here, for each k = 1, 2, ..., n, $\rho_{1k}$, $\rho_{2k}$, ..., $\rho_{nk}$ are the components of the vector $\mathbf{e}_k$ in the basis **f**. Expressing $\mathbf{e}_k$ in (1.6) in terms of the $\mathbf{f}_j$s by means of (1.7) gives

$$\sum_{k=1}^{n} \xi_k \left( \sum_{j=1}^{n} \rho_{jk} \mathbf{f}_j \right) = \sum_{k=1}^{n} \eta_k\, \mathbf{f}_k = \sum_{j=1}^{n} \eta_j\, \mathbf{f}_j; \tag{1.8}$$

the last equality comes about because of the purely cosmetic step of replacing the summation index k by j in the central expression in (1.8). Since the **f**s are linearly independent, the coefficients of $\mathbf{f}_j$ in the extreme members of (1.8) must coincide, yielding

$$\eta_j = \sum_{k=1}^{n} \rho_{jk}\, \xi_k,\ j = 1, ..., n. \tag{1.9}$$

This "change-of-basis" formula expresses the components $\eta_j$ of **x** in the basis **f** in terms of the components $\xi_k$ of **x** in the basis **e** with the help of the $n^2$ scalars $\rho_{jk}$ that determine the relation (1.7) between the two bases.

Calculations with components of vectors often lead to repulsive expressions like those appearing in (1.8), or worse. There is a notational abbreviation that helps. Inspection of (1.6)–(1.9) reveals that the subscript subject to summation always occurs twice in the summand. This suggests that we *delete* the summation sign Σ from the equations, and agree to sum *automatically* over the repeated index. With this *summation convention* in force, (1.9) would be replaced by

$$\eta_j = \rho_{jk}\, \xi_k,\ j = 1, ..., n. \tag{1.10}$$

To make this foolproof, we must understand that the summation automatically extends over the *range* 1, 2, ..., n of possible values of the subscripts. In both (1.9) and (1.10), the subscripts j and k are different in character: the index k that is repeated on one side of the equation is "summed out" and could be replaced by any other symbol (m or p, for example), while the subscript j—called the "free index"—takes successively the values indicated at the end of the equation. We can extend our notational agreement by adopting the *range convention*: the free index is *assumed* to take the values 1, ..., n unless otherwise indicated; we then omit the range of j in (1.10) and replace the statement by

$$\eta_j = \rho_{jk}\, \xi_k. \tag{1.11}$$

Note that the same free index appears on each side of the equation (1.11), while the repeated index does not. Under the new notational agreements, we would write (1.8), for example, as

$$\rho_{jk}\, \xi_k\, \mathbf{f}_j = \eta_k\, \mathbf{f}_k = \eta_j\, \mathbf{f}_j, \tag{1.12}$$

with *two* summations implied in the left-most member.

We shall assume the summation and range conventions to be in force in what follows. There will, however, be some embarassing moments when the summation convention gets us into some difficulty; when this happens, we shall explicitly suspend the convention temporarily.

Let R be a vector space of finite or infinite dimension. Let M be a subset of R. M is said to be a *linear manifold* in R if M itself is a vector space under the operations of addition and multiplication by a scalar that apply in the parent space R. M may or may not be finite dimensional. In finite dimensional spaces, the term *subspace* is often used instead of linear manifold.

In the vector space A of arrows in the plane, let **e** be a fixed arrow, and let M be the set of all arrows of the form $\alpha\mathbf{e}$, with $\alpha$ ranging over all the real numbers. Clearly M contains the sum of any two arrows of this form, as well as every scalar multiple of such an arrow. Thus M is a linear manifold in A. As a vector space in its own right, the dimension of M is one.

Consider the space $R_3$ of columns of three real numbers $x_1$, $x_2$, $x_3$; let us think of $R_3$ as the collection of coordinates of points in ordinary physical space. Let M be the set of all columns for which $\alpha_1 x_1 + \alpha_2 x_2 + \alpha_3 x_3 = 0$, where $\alpha_1$, $\alpha_2$, $\alpha_3$ are fixed scalars. It is easily verified that M is a two-dimensional linear manifold in $R_3$. Geometrically, it represents the set of coordinates of all points that lie in the plane through the origin that is normal to the line described in parametric form by $x_1 = \alpha_1\tau$, $x_2 = \alpha_2\tau$, $x_3 = \alpha_3\tau$, $-\infty < \tau < \infty$.

The space of solutions on $[0, \pi]$ of the ordinary differential equation (1.2) is a two-dimensional manifold in the space C of continuous real-valued functions on $[0, \pi]$; the space $P_m$ of polynomials is a linear manifold in C as well. The collection $C^{(1)}$ of all real-valued *continuously differentiable* functions on $[0, \pi]$ is an infinite dimensional vector space that is also a linear manifold in C; this is the subject of Problem 1.10.

Let R be any vector space. Let $\mathbf{x}_1$, $\mathbf{x}_2$, ..., $\mathbf{x}_k$ be k linearly independent vectors in R. Let M be the set of all vectors of the form $\alpha_1\mathbf{x}_1 + \alpha_2\mathbf{x}_2 + \cdots + \alpha_k\mathbf{x}_k$, where the $\alpha$s take all possible scalar values. The k-dimensional linear manifold M is called the *span* of the k vectors $\mathbf{x}_j$.

The landscape in an abstract vector space R is as yet rather barren. To endow R with some geometric structure, and thus make it much more interesting and useful, we turn to the notion of Euclidean (or "inner product") spaces.

Let R be any *real* vector space. We wish to introduce into R the metric notions of length and angle familiar from our experience in everyday physical space. You may recall that, when the arrows were first inflicted upon you, you were told that the "dot" product of two arrows **x** and **y** was the product of the lengths $|\mathbf{x}|$ and $|\mathbf{y}|$ of the arrows and the cosine of the angle $\theta$ between them; in symbols,

$$\mathbf{x} \cdot \mathbf{y} = |\mathbf{x}|\,|\mathbf{y}| \cos \theta, \tag{1.13}$$

at least when **x** and **y** were non-null arrows. From (1.13), it follows incidentally that

$$|\mathbf{x}| = (\mathbf{x} \cdot \mathbf{x})^{1/2}. \tag{1.14}$$

The definition of dot product given above clearly relates to the geometrical notion of projection of one arrow on another. It is easy to verify from (1.13) that the dot product of two arrows has the following properties:

$$\mathbf{x} \cdot \mathbf{y} = \mathbf{y} \cdot \mathbf{x}, \tag{1.15}$$

$$(\mathbf{x} + \mathbf{y}) \cdot \mathbf{z} = \mathbf{x} \cdot \mathbf{z} + \mathbf{y} \cdot \mathbf{z}, \tag{1.16}$$

$$(\alpha\mathbf{x}) \cdot \mathbf{y} = \alpha(\mathbf{x} \cdot \mathbf{y}), \tag{1.17}$$

$$\mathbf{x} \cdot \mathbf{x} > 0 \text{ if } \mathbf{x} \neq \mathbf{o}; \tag{1.18}$$

here **x**, **y**, and **z** are arbitrary arrows, and $\alpha$ is a scalar, necessarily a real number in the present case.

It turns out that the characteristics of the dot product listed in (1.15)–(1.18) are the only crucial ones for the kind of metric calculations we shall wish to make. This suggests a way to introduce the notion of dot product—or *scalar* product, as we prefer to call it generally—in an *arbitrary* real vector space R: let a rule be given which assigns to every *pair* of vectors **x**, **y** in R a scalar (i.e., a real number) denoted by (**x**, **y**). Suppose this rule conforms to (1.15)–(1.18), where it is understood that $\mathbf{x} \cdot \mathbf{y}$ is now replaced by (**x**, **y**). Then we shall say that (**x**, **y**) is a scalar product for R. A real vector space equipped with a scalar product is called a *real Euclidean space*. (There is an implicit question lurking in the shadows of this discussion: in a given real vector space, can one *really* assign two or more rules for forming a scalar product, each obeying the basic requirements (1.15)–(1.18)? This issue is explored in Problem 1.12.)

Given a real vector space R and a scalar product (**x**, **y**), we define the length of any vector $\mathbf{x}$ in R by $|\mathbf{x}| = (\mathbf{x}, \mathbf{x})^{1/2}$, having (1.14) in mind.

The rules (1.15)–(1.18) can be used to show that the null vector $\mathbf{o}$, and only the null vector, has zero length, and that $(\mathbf{o}, \mathbf{x}) = 0$ for every $\mathbf{x}$ in R. It is also true that the only vector $\mathbf{y}$ in R such that $(\mathbf{y}, \mathbf{x}) = 0$ for every $\mathbf{x}$ in R is $\mathbf{y} = \mathbf{o}$.

We now show that, no matter how the scalar product $(\mathbf{x}, \mathbf{y})$ is defined in detail on R, it is always possible to make use of it to associate with any pair of non-null vectors $\mathbf{x}$ and $\mathbf{y}$ an "angle $\theta$ between them" such that the counterpart of (1.13) for arrows holds: $(\mathbf{x}, \mathbf{y}) = |\mathbf{x}||\mathbf{y}| \cos \theta$. To do this, we first need to establish the *Schwarz inequality* in an arbitrary real Euclidean space: for every $\mathbf{x}$, $\mathbf{y}$ in R,

$$|(\mathbf{x}, \mathbf{y})| \leq |\mathbf{x}||\mathbf{y}|. \tag{1.19}$$

In view of the remarks in the two preceding paragraphs, (1.19) certainly holds when either $\mathbf{x}$ or $\mathbf{y}$ is the null vector. To demonstrate (1.19) for *all* $\mathbf{x}$, $\mathbf{y}$ in R, we need therefore only deal with the case in which neither $\mathbf{x}$ nor $\mathbf{y}$ is $\mathbf{o}$. To this end, we use the following device: for fixed $\mathbf{x} \neq \mathbf{o}$ and $\mathbf{y} \neq \mathbf{o}$ and for every real number t, define $f(t) = |\mathbf{x} + t\,\mathbf{y}|^2$, and observe that $f(t) \geq 0$ for every t. Using (1.15)–(1.18), one finds the alternate representation $f(t) = |\mathbf{x}|^2 + 2\,(\mathbf{x}, \mathbf{y})t + |\mathbf{y}|^2 t^2$. A plot of f(t) vs. t reveals a parabola whose minimum is necessarily on or above the t-axis. This means the quadratic expression for f(t) must have a non-positive discriminant: $4[(\mathbf{x}, \mathbf{y})^2 - |\mathbf{x}|^2\,|\mathbf{y}|^2] \leq 0$. But this is equivalent to (1.19), completing the proof of the Schwarz inequality.

In (1.19), equality *might* hold. As discussed in Problem 1.14, this occurs if and only if $\mathbf{x}$ and $\mathbf{y}$ are linearly dependent.

For $\mathbf{x} \neq \mathbf{o}$, $\mathbf{y} \neq \mathbf{o}$, the Schwarz inequality (1.19) asserts that $|(\mathbf{x}, \mathbf{y})||\mathbf{x}|^{-1}|\mathbf{y}|^{-1} \leq 1$, so that there is an angle $\theta$ such that $(\mathbf{x}, \mathbf{y}) = |\mathbf{x}||\mathbf{y}| \cos \theta$. Thus, even in an *arbitrary* real Euclidean space, we can interpret the scalar product geometrically as yielding the projection of one vector upon the other, multiplied by the length of the second vector.

The Schwarz inequality (1.19) implies the "triangle" inequality: for any two vectors $\mathbf{x}$ and $\mathbf{y}$, the inequality $|\mathbf{x} + \mathbf{y}| \leq |\mathbf{x}| + |\mathbf{y}|$ must hold; see Problem 1.15.

Two vectors $\mathbf{x}$ and $\mathbf{y}$ in a real Euclidean space R are *orthogonal* if $(\mathbf{x}, \mathbf{y}) = 0$; if neither $\mathbf{x}$ nor $\mathbf{y}$ is null, then the angle $\theta$ between them is $\pi/2$. If $\mathbf{x}$ and $\mathbf{y}$ are orthogonal, the Pythagorean theorem holds: $|\mathbf{x} + \mathbf{y}|^2 = |\mathbf{x}|^2 + |\mathbf{y}|^2$. Only the null vector is orthogonal to every vector in R. These facts follow easily from the rules (1.15)–(1.18) for scalar products.

Let R be a real Euclidean space whose dimension is finite, say dim R = n. Let $\mathbf{e}_1, \ldots, \mathbf{e}_n$ be basis for R. If $(\mathbf{e}_i, \mathbf{e}_j) = 0$ whenever i and j are distinct, the basis is said to be orthogonal. If in addition the length of each basis vector is unity, the basis is *orthonormal*. Thus for an orthonormal basis,

$$(\mathbf{e}_i, \mathbf{e}_j) = \delta_{ij}, \tag{1.20}$$

where $\delta_{ij}$ is the Kronecker delta; $\delta_{ij} = 1$ if i and j are the same, $\delta_{ij} = 0$ otherwise. Clearly, there is a question: in a finite-dimensional real Euclidean space, is it necessary that an orthonormal basis exists? As shown in Problem 1.17, there is an algorithm—called the Gram-Schmidt process after its inventors—that will manufacture an orthonormal basis out of the elements of an *arbitrarily selected* basis for R. Since any finite-dimensional vector space has a basis, the Gram-Schmidt process not only assures us that an orthonormal basis exists in every finite-dimensional real Euclidean space, but it also provides a method for constructing one. An example of such a construction in a space of useful polynomials may be found in Problem 1.18.

Let $\mathbf{e}_1, \mathbf{e}_2, ..., \mathbf{e}_n$ be an orthonormal basis for a real Euclidean space R, and let $\mathbf{x}$ be any vector in R. Since the **e**s form a basis, we can write $\mathbf{x} = \xi_i\, \mathbf{e}_i$ (summation convention), where the scalars $\xi_i$ are the components of $\mathbf{x}$ in the basis **e**. The orthonormality of **e** makes it possible to give a simple formula for these components with the help of the scalar product: $\xi_i = (\mathbf{x}, \mathbf{e}_i)$. We thus have the following elegant and useful representation for any $\mathbf{x}$ in R:

$$\mathbf{x} = (\mathbf{x}, \mathbf{e}_i)\, \mathbf{e}_i; \tag{1.21}$$

here summation on the repeated index i is understood.

***Example 1.6.*** *The set of column vectors as a real Euclidean space.* Let $\mathsf{R} = \mathsf{R}_n$ be the real n-dimensional vector space of columns of the form (1.1) of Example 1.1. Define a candidate for scalar product on $\mathsf{R}_n$ by

$$(\mathbf{x}, \mathbf{y}) = x_i\, y_i, \tag{1.22}$$

where $x_i$ and $y_i$ are the elements of the column vectors $\mathbf{x}$ and $\mathbf{y}$, respectively, and the summation and range conventions introduced earlier are in force. One can now verify that each of the requirements (1.15)–(1.18) necessary to qualify $(\mathbf{x}, \mathbf{y})$ as a legitimate scalar product is indeed satisfied. This is the most commonly used scalar product in this particular vector space, and indeed when we speak of the *real Euclidean space* $\mathsf{R}_n$, we shall take for granted that the scalar product is that of (1.22). As shown in Problem 1.12, though, there are other possible choices of scalar product that may be used to render $\mathsf{R}_n$ a real Euclidean space.

For the real Euclidean space $\mathsf{R}_n$, the Schwarz inequality (1.19) asserts that

$$|x_i\, y_i| \leq (x_i\, x_i)^{1/2}\, (y_j\, y_j)^{1/2}; \tag{1.23}$$

remember the summation convention. The inequality (1.23) for real numbers is one with which you may be familiar.

Let $\mathbf{e}_1, \mathbf{e}_2, ..., \mathbf{e}_n$ be the n vectors in $\mathsf{R}_n$ for which the column $\mathbf{e}_i$ has all entries zero except the $i^{th}$, which is one. These vectors certainly comprise an orthonormal basis for $\mathsf{R}_n$. (It is easy to see that there are other orthonormal bases for $\mathsf{R}_n$). You might wish to verify that (1.21) holds for these $\mathbf{e}_i$s.

***Example 1.7.*** *The continuous functions as a real Euclidean space.* Consider now the infinite-dimensional vector space C of continuous real-valued functions on $[0, \pi]$ discussed in Example 1.3. A few calculations, accompanied by a little reflection, will verify that the following proposal for a scalar product works: for any pair of functions $\varphi$ and $\psi$ in C, set

$$(\varphi, \psi) = \int_0^{\pi} \varphi(t)\ \psi(t)\ dt. \tag{1.24}$$

Again, we take for granted that the scalar product is that of (1.24) when we refer to C as a real Euclidean space. In this case, the "length" of a function $\varphi$ in C is

$$|\varphi| = \left(\int_0^{\pi} \varphi^2(t)\ dt\right)^{1/2}, \tag{1.25}$$

and the "angle" between two such functions is determined by

$$\cos\theta = \frac{\int_0^{\pi} \varphi(t)\ \psi(t)\ dt}{\left[\int_0^{\pi} \varphi^2(t)\ dt\right]^{1/2}\left[\int_0^{\pi} \psi^2(t)\ dt\right]^{1/2}}. \tag{1.26}$$

Since C is an infinite-dimensional space, we cannot speak of bases, orthonormal or otherwise, for C. So we can't find the counterpart in C of the representation theorem (1.21). This state of affairs seems quite unsatisfying, and almost cries out to be investigated. To do so is to open the door to a new and beautiful room; we won't do it here because there are too many difficult technical questions, but the reader who wishes to peek through this door might like to consult references [1.1] and [1.4–1.6] at the end of this chapter. A fascinating discovery that one makes upon exploring this new room is that C is not really quite the right vector space in which to be asking these questions; *finding* the right space is part of the adventure.

***Example 1.8.*** *The trigonometric polynomials.* Let N be a positive integer, and set $n = 2N + 1$. Let $\mathsf{F}_n$ be the set of all real valued functions $\varphi$ of the form

$$\varphi(t) = \alpha_0 + \sum_{k=1}^{N} (\alpha_k \cos kt + \beta_k \sin kt), \tag{1.27}$$

where the $\alpha_k$s and $\beta_k$s are real numbers; observe that every such $\varphi$ is periodic with period $2\pi$. $\mathsf{F}_n$ is a vector space under the natural operations of addition and multiplication by a scalar; the dimension of $\mathsf{F}_n$ is n. Let the scalar product of two "trigonometric polynomials" $\varphi$ and $\psi$ of the form (1.27) be given by

$$(\varphi, \psi) = \int_0^{2\pi} \varphi(t)\, \psi(t)\, dt. \tag{1.28}$$

Define the n functions $\chi_j(t)$ as follows

$$\chi_j(t) = \begin{cases} (2\pi)^{-1/2} \text{ for } j = 1, \\ \pi^{-1/2} \cos (j - 1)t \text{ for } j = 2, ..., N + 1, \\ \pi^{-1/2} \sin (j - N - 1)t \text{ for } j = N + 2, ..., 2N + 1. \end{cases} \tag{1.29}$$

One can check that $(\chi_i, \chi_j) = \delta_{ij}$, so that $\chi_1, ..., \chi_n$ form an orthonormal basis: every $\varphi$ in $\mathsf{F}_n$ can be represented in the form

$$\varphi = (\varphi, \chi_j)\chi_j, \tag{1.30}$$

where according to the range convention, the sum in (1.30) extends from $j = 1$ to $j = n = 2N + 1$.

If we were to *formally* generalize (1.29) by including *all* of the infinitely many cosines and sines with integer circular frequency, the right side of (1.30) would represent a formal *Fourier series* for a function $\varphi$. This raises the question of whether, in an appropriate infinite-dimensional vector space of real-valued functions, there might be a sense in which the functions $(2\pi)^{-1/2}$, $\pi^{-1/2} \cos t$, $\pi^{-1/2} \cos 2t$, ..., $\pi^{-1/2} \sin t$, $\pi^{-1/2}\sin 2t$, ... would form an orthonormal basis when the scalar product is chosen as in (1.28). This question is related to the issue left hanging at the end of the preceding example; the reader is referred to the books cited there for further information.

What happens if the vector space $\mathsf{R}$ has complex scalars, as in Example 1.2? One can invent scalar products on complex vector spaces, too. Such a scalar product assigns to each pair of vectors **x**, **y** in the space a scalar (**x**, **y**), now a *complex* number, in accordance with the following rules that take the place of (1.15)–(1.18):

$$(\mathbf{x}, \mathbf{y}) = \overline{(\mathbf{y}, \mathbf{x})}, \tag{1.31}$$

$$(\mathbf{x} + \mathbf{y}, \mathbf{z}) = (\mathbf{x}, \mathbf{z}) + (\mathbf{y}, \mathbf{z}), \tag{1.32}$$

$$(\alpha\mathbf{x}, \mathbf{y}) = \alpha(\mathbf{x}, \mathbf{y}), \tag{1.33}$$

$$(\mathbf{x}, \mathbf{x}) > 0 \text{ if } \mathbf{x} \neq \mathbf{o}.; \tag{1.34}$$

The overbar in (1.31) indicates the complex conjugate of the number $(\mathbf{y}, \mathbf{x})$. Note that (1.31) implies that $(\mathbf{x}, \mathbf{x})$ is always real, and (1.33) with $\alpha = 0$ asserts that $(\mathbf{o}, \mathbf{y}) = 0$ for every $\mathbf{y}$ in $\mathsf{R}$, so that in particular, $(\mathbf{o}, \mathbf{o}) = 0$. Naturally, the length of a vector $\mathbf{x}$ in the complex space $\mathsf{R}$ is defined to be $|\mathbf{x}| = (\mathbf{x}, \mathbf{x})^{1/2}$.

In texts aimed at mathematicians, it is usually the *complex Euclidean space* (or *unitary* space, as it is often called) that is the principal subject of study. This is the case, for example, in the book by Halmos [1.3] cited at the end of this chapter. Since our main concern is with the structure of real spaces and their application, we leave the issue of complex spaces, too, to the reader to pursue. We touch on complex spaces only where we wish to contrast the nature of the results in such spaces with those in spaces whose scalars are real. The differences between the two cases are rarely emphasized, and even more rarely understood by students who have been through the usual courses in linear algebra. The book by Gel'fand [1.2] in the list below has an unusually clear discussion of *real* Euclidean spaces; see pp. 114–26 of [1.2].

## References

[1.1] L. Debnath and K. Mikusinski, *Introduction to Hilbert Spaces with Applications*, Academic Press, New York, 1990.

[1.2] I.M. Gel'fand, *Lectures on Linear Algebra*, translated from the Russian by A. Shenitzer, Dover Publications, New York, 1989.

[1.3] P.R. Halmos, *Finite Dimensional Vector Spaces*, Second Edition, Van Nostrand-Reinhold, New York, 1958.

[1.4] P.R. Halmos, *Introduction to Hilbert Space*, Chelsea Publishing Company, New York, 1951.

[1.5] A.N. Kolmogorov and S.V. Fomin, *Elements of the Theory of Functions and Functional Analysis*, Vol. 1: *Metric and Normed Spaces*, Vol. 2: *Measure. The Lebesgue Integral. Hilbert Space*. Graylock Press, Rochester, New York, 1951.

[1.6] A.W. Naylor and G.R. Sell, *Linear Operator Theory in Engineering and Science*, Springer-Verlag, New York, 1982.

## Problems

**1.1.** Example 1.4 concerns the vector space of real solutions of the differential equation (1.2) on the interval $[0, \pi]$. Suppose that (1.2) were replaced by the differential equation $\varphi'' + \varphi = 1$. Would the set of all solutions on $[0, \pi]$ of this new differential equation comprise a vector space?

**1.2.** In Example 1.5, the vector space under discussion consists of all real polynomials of degree *not exceeding* N. Would the set of real polynomials whose degree is *exactly* N comprise a vector space?

**1.3.** Let $\mathsf{R}$ be any vector space, and let $\{\mathbf{x}_1, \mathbf{x}_2, ..., \mathbf{x}_k\}$ be a set of k vectors, one of which is the null vector. Show that the set of vectors is linearly dependent.

**1.4.** Show geometrically that any set of three or more arrows in the set $\mathsf{A}_2$ of arrows in a plane is a linearly dependent set.

**1.5.** Show that any set of three solutions of the differential equation (1.2) is a linearly dependent set.

**1.6.** Show that, for any positive integer k, the set of real-valued functions $\varphi_0(t), \varphi_1(t), ..., \varphi_k(t)$ defined by $\varphi_j(t) = t^j$, $j = 0, 1, ..., k$, constitutes a linearly independent set in the space $\mathsf{C}$ of continuous functions on $[0, \pi]$.

**1.7.** Let $\mathsf{R}$ be a finite dimensional vector space, and let $\mathbf{e}_1, ..., \mathbf{e}_n$ be a basis in $\mathsf{R}$. Show that the components in $\mathbf{e}$ of a vector in $\mathsf{R}$ are unique.

**1.8.** Consider the basis $\mathbf{p}_1(t) = 1$, $\mathbf{p}_2(t) = t$, $\mathbf{p}_3 = t^2$, ..., $\mathbf{p}_n(t) = t^{n-1}$ in the vector space $\mathsf{P}_n$ of real polynomials of degree $N = n - 1$. Show that the components $\xi_1, \xi_2, ..., \xi_n$ in this basis of an arbitrary polynomial $\mathbf{p}(t)$ in $\mathsf{P}_n$ are given by $\xi_j = \mathbf{p}^{(j-1)}(0)/(j-1)!, j = 1, ..., n$, where $\mathbf{p}^{(j)}(t)$ stands for the $j^{th}$ derivative of $\mathbf{p}(t)$.

**1.9.** Let $\mathsf{R}$ be a vector space in which the vectors are complex numbers. (a) If the scalars are real numbers, find the dimension of $\mathsf{R}$ and a basis. (b) Answer the same questions if the scalars are *complex* numbers.

**1.10.** Show that the set $\mathsf{C}^{(1)}$ of continuously differentiable functions on $[0, \pi]$ is a linear manifold in the vector space $\mathsf{C}$ of continuous real-valued functions on $[0, \pi]$.

**1.11.** Let $\mathsf{R}$ be the set of all $3 \times 3$ matrices of real numbers $a_{ij}$ with real scalars and under the natural operations of addition and multiplication by a scalar. Let $\mathsf{R}'$ be the subset of $\mathsf{R}$ consisting of all skew-symmetric matrices, i.e., all matrices for which $a_{ij} = -a_{ji}$. Show that $\mathsf{R}'$ is a linear manifold in $\mathsf{R}$, find its dimension and determine a basis for it.

**1.12.** Consider the two-dimensional real vector space $\mathsf{R}_2$ of columns of two real numbers. Show that $(\mathbf{x}, \mathbf{y}) = x_1 y_1 + x_2 y_2$ and $(\mathbf{x}, \mathbf{y}) = x_1y_1 +$

$(1/2)(x_1 y_2 + x_2 y_1) + x_2 y_2$ are both legitimate scalar products in $R_2$. Find the angle between $\binom{1}{0}$ and $\binom{0}{1}$ for each of these scalar products.

**1.13.** Let R be any real Euclidean space. Show that any finite orthonormal set of vectors is linearly independent.

**1.14.** In the Schwarz inequality (1.19), show that equality holds if and only if **x** and **y** are linearly dependent.

**1.15.** Show that the Schwarz inequality (1.19) implies the *triangle inequality*: $|\mathbf{x} + \mathbf{y}| \leq |\mathbf{x}| + |\mathbf{y}|$.

**1.16.** Let C be the real Euclidean space of all continuous, real-valued functions on $[0, \pi]$ of Example 1.7, the scalar product being that introduced in (1.24). Find the angle between the functions cos kt and sin kt, where k is a positive integer.

**1.17.** *The Gram-Schmidt process.* Let R be any real Euclidean space, and let $\mathbf{f}_1, \mathbf{f}_2, ..., \mathbf{f}_k$ be k linearly independent vectors in R. Define k vectors $\mathbf{e}_1, \mathbf{e}_2, ..., \mathbf{e}_k$ recursively as follows:

$$\mathbf{e}_1 = \mathbf{f}_1,\ \mathbf{e}_2 = \mathbf{f}_2 - \frac{(\mathbf{f}_2, \mathbf{e}_1)}{(\mathbf{e}_1, \mathbf{e}_1)}\mathbf{e}_1,\ ...,\ \mathbf{e}_j = \mathbf{f}_j - \sum_{i=1}^{j-1} \frac{(\mathbf{f}_j, \mathbf{e}_i)}{(\mathbf{e}_i, \mathbf{e}_i)}\mathbf{e}_i$$

$$j = 2, 3, ...\ k.$$

(a) Show that $\mathbf{e}_1, \mathbf{e}_2, ..., \mathbf{e}_k$ form an orthogonal set. (b) Give a geometric interpretation of the algorithm described above in the setting of the vector space $A_2$ of arrows in the plane. Observe that if R is finite dimensional with dimension k, then $\mathbf{e}_1, \mathbf{e}_2, ..., \mathbf{e}_k$ form an orthogonal basis for R.

**1.18.** The polynomials $\mathbf{p}_0 = 1$, $\mathbf{p}_1 = t$, $\mathbf{p}_2 = t^2$ and $\mathbf{p}_3 = t^3$ form a basis for the space $P_4$ of real polynomials of degree not exceeding three. Taking the scalar product of two polynomials **p** and **q** to be

$$(\mathbf{p}, \mathbf{q}) = \int_{-1}^{1} \mathbf{p}(t)\, \mathbf{q}(t)\, dt,$$

use the Gram-Schmidt procedure of the preceding problem to manufacture an orthogonal basis from the $\mathbf{p}_i$s. You might wish to investigate the relationship between the orthogonal basis for $P_4$ constructed in this way and the first few Legendre polynomials.

**1.19.** Let R be the vector space consisting of all $n \times n$ matrices of real numbers, with real scalars. Propose a scalar product on R as follows: let $\underline{A} = (a_{ij})$ and $\underline{B} = (b_{ij})$ be two matrices in R. Set $(\underline{A}, \underline{B}) = a_{ij}b_{ij}$, bearing in mind the summation convention. Is this a legitimate scalar product on R?

**1.20.** Let R be the set of all infinite sequences $\mathbf{x} = \{x_1, x_2, x_3, ...\}$ of real numbers with the property that the infinite series $\sum_{i=1}^{\infty} x_i^2$ converges. (a) Does R, along with the real numbers as scalars, comprise a vector space? (b) Consider the set $E_N$ of N elements $\mathbf{e}_i$ of R, i = 1, ..., N, where $\mathbf{e}_i = \{0, 0, ..., 1, 0, ...\}$, with one in the $i^{th}$ place, zeros elsewhere. Show that $E_N$ is a linearly independent set for every positive integer N. What is the dimension of R? (c) If $\mathbf{x} = \{x_1, x_2, ...\}$ and $\mathbf{y} = \{y_1,, y_2, ...\}$ are two elements of R, set $(\mathbf{x}, \mathbf{y}) = \sum_{i=1}^{\infty} x_i y_i$; show that $(\mathbf{x}, \mathbf{y})$ is a proper scalar product for R. The space R is usually denoted by $l_2$.

CHAPTER 2

# LINEAR TRANSFORMATIONS

Vector spaces would be rather boring places without the notion of *linear transformation*, which we are about to describe.

## Linear Transformations

Let R be a real *or* complex vector space. For us, a transformation **A** is a mapping that assigns to each vector **x** in R another vector **y** in R: we write

$$\mathbf{y} = \mathbf{A}(\mathbf{x}) = \mathbf{A}\mathbf{x}; \tag{2.1}$$

the parentheses in (2.1) may be either included or omitted in the equations that follow, depending on whether they are needed for clarity. We call **y** the *image* of **x** under **A**. The only requirements imposed on **A** necessary to qualify it as a transformation are that its input **x** be in R, its output **y** be in R, and its output consist of only *one* vector, rather than several, so that **A** is single-valued. A transformation **A** is *linear* if it has the following two special properties:

$$\mathbf{A}(\mathbf{x} + \mathbf{y}) = \mathbf{A}(\mathbf{x}) + \mathbf{A}(\mathbf{y}) \text{ for every } \mathbf{x}, \mathbf{y} \text{ in R}, \tag{2.2}$$

$$\mathbf{A}(\alpha\mathbf{x}) = \alpha\, \mathbf{A}(\mathbf{x}) \text{ for every scalar } \alpha \text{ and every } \mathbf{x} \text{ in R}. \tag{2.3}$$

Although *nonlinear* transformations are of great interest and are now pursued in mathematical research and in physical applications with great intensity, linear transformations are much easier to understand, and we limit our study to them exclusively. Fortunately, they have an enormous range of applicability in the physical sciences, in engineering, and elsewhere.

***Example 2.1.*** *Linear transformations on* $R_2$. Let the vector space be $R_2$, consisting of columns of two real elements, the scalars being real; see Chapter 1, Example 1.1. Let **x** and **y** be two column vectors in $R_2$ related by

$$\mathbf{y} = \begin{pmatrix} y_1 \\ y_2 \end{pmatrix} = \begin{pmatrix} a_{11} & a_{12} \\ a_{21} & a_{22} \end{pmatrix} \begin{pmatrix} x_1 \\ x_2 \end{pmatrix} = \begin{pmatrix} a_{11}x_1 + a_{12}x_2 \\ a_{21}x_1 + a_{22}x_2 \end{pmatrix} = \mathbf{Ax}, \qquad (2.4)$$

where the $a_{ij}$s are four given real numbers forming the elements of the 2 × 2 matrix shown. Thus the input to the transformation **A** is **x**, the output **y**. One shows easily that **A** is a linear transformation. This example has an obvious generalization to the real space $\mathsf{R}_n$ of columns of n real elements, where n is any positive integer.

***Example 2.2.*** *A linear transformation on the space* $\mathsf{P}_n$ *of polynomials of degree not greater than* $N = n - 1$. Let $\mathbf{p}(t) = \alpha_0 + \alpha_1 t + \alpha_2 t^2 + ... + \alpha_N t^N$ be a polynomial in the space $\mathsf{P}_n$ of Example 1.5. Let **A** be the transformation that takes **p** into the polynomial **q** in $\mathsf{P}_n$, where q is defined by

$$\mathbf{q}(t) = \mathbf{p}'(t) = \alpha_1 + 2\alpha_2 t + ... + N\alpha_N t^{N-1} = \mathbf{Ap}(t), \qquad (2.5)$$

and the prime indicates differentiation with respect to t. **A** is clearly a linear transformation on $\mathsf{P}_n$. Let **B** be a second linear transformation on $\mathsf{P}_n$ defined by

$$\mathbf{q}(t) = \mathbf{Bp}(t) = \frac{1}{t}\int_0^t \mathbf{p}(\tau)\, d\tau. \qquad (2.6)$$

(Be sure you see that **q** in (2.6) is indeed in $\mathsf{P}_n$; note that without the factor 1/t in front of the integral, this would not be generally true.) For future purposes, we raise the following question: suppose we apply **A** to **p**, then apply **B** to the resulting output. Does the final output differ from that obtained if we first apply **B** to p, *then* apply **A**? Calculation shows that

$$\mathbf{B}(\mathbf{Ap}) = \alpha_1 + \alpha_2 t + ... + \alpha_N t^{N-1},$$

$$\mathbf{A}(\mathbf{Bp}) = \frac{\alpha_1}{2} + \frac{2\alpha_2}{3} t + ... + \frac{N\alpha_N}{N+1} t^{N-1}, \qquad (2.7)$$

so that indeed, $\mathbf{B}(\mathbf{Ap}) \neq \mathbf{A}(\mathbf{Bp})$, indicating that the order in which linear transformations are applied may be important.

***Example 2.3.*** *An integral operator on* $\mathsf{C}$. Let $\mathsf{C}$ be the space of continuous real-valued functions on [0, π]. With any function f in $\mathsf{C}$, let us associate another function g = **A**f defined by

$$g(t) = \int_0^\pi \cos(t - \tau)\, f(\tau)\, d\tau. \qquad (2.8)$$

It is clear that g is also in $\mathsf{C}$, and that the rule **A** that assigns g to f is indeed a linear transformation of $\mathsf{C}$ into itself. This sort of relationship, usually with $\cos(t - \tau)$ replaced by some more complicated function of t and $\tau$, sometimes arises in applications in circumstances such that g(t) is "known" in terms of given information in the problem, while f(t) is to be found. One then refers to (2.8) as an *integral equation* for f; indeed, it is an example of an *integral equation of the first kind of Fredholm type*. For an introductory discussion of integral equations, the reader may consult Chapter 3 of the book by R. Courant and D. Hilbert [2.2] cited at the end of this chapter.

We shall encounter many more examples of linear transformations as we proceed. Let us note here a final example for the moment, or rather what might better be called a *non-example*. Let C be as in the preceding example, and define a transformation **A** by $\mathbf{A}\varphi = \varphi^2$ for every $\varphi$ in C. Clearly $\varphi^2$ is continuous because $\varphi$ is, so indeed **A** in a transformation of C into itself, but **A** is certainly not linear.

Two linear transformations **A** and **B** are said to be equal if, for every common input, they have a common output: $\mathbf{Ax} = \mathbf{Bx}$ for every **x** in R. The *sum* of two linear transformations $\mathbf{A} + \mathbf{B}$ is the transformation defined by $(\mathbf{A} + \mathbf{B})\mathbf{x} = \mathbf{Ax} + \mathbf{Bx}$, and by the scalar multiple $\alpha\mathbf{A}$ of a linear transformation, we mean $(\alpha\mathbf{A})\mathbf{x} = \alpha(\mathbf{Ax})$. The *null* linear transformation **O** is the one for which $\mathbf{Ox} = \mathbf{o}$ for all **x**; it has the property that $\mathbf{A} + \mathbf{O} = \mathbf{O} + \mathbf{A} = \mathbf{A}$. Indeed, it is easy to verify the following assertion, which will prove to be of interest to us later: *the collection* L *of all linear transformations on a vector space* R *is itself a vector space under the operations of addition and multiplication by a scalar as defined above.*

Beyond this, we can also speak of the *product* of two linear transformations: **AB** is defined by $(\mathbf{AB})\mathbf{x} = \mathbf{A}(\mathbf{Bx})$. The discussion in Example 2.2 makes clear the troublesome fact that, in general, $\mathbf{AB} \neq \mathbf{BA}$, so that this kind of multiplication is not commutative. Nevertheless, the *identity* transformation **1**, defined by $\mathbf{1x} = \mathbf{x}$ for every **x** in R, has the property that $\mathbf{1A} = \mathbf{A1} = \mathbf{A}$ for any **A**, and thus commutes with any linear transformation.

Let **A** be a linear transformation on a vector space R. The *null space*, or *kernel*, of **A**—denoted by $N(\mathbf{A})$—is the set of all vectors **x** in R such that $\mathbf{Ax} = \mathbf{0}$. The *range*, or *image*, of **A** is the set $R(\mathbf{A})$ of all possible outputs of **A**: $R(\mathbf{A}) = \{\mathbf{y} \in \mathsf{R} | \mathbf{y} = \mathbf{Ax}$ for at least one **x** in R$\}$. It is obvious that $R(\mathbf{1}) = \mathsf{R}$, $N(\mathbf{1}) = \{\mathbf{o}\}$, $R(\mathbf{O}) = \{\mathbf{o}\}$, $N(\mathbf{O}) = \mathsf{R}$; here we have used the symbol $\{\mathbf{o}\}$ to designate the set consisting of the null vector alone. It is easy to see that, for *any* linear transformation, both $R(\mathbf{A})$ and $N(\mathbf{A})$ are linear manifolds in R.

***Example 2.4.*** *Range and null space of a linear transformation in* $\mathsf{R}_2$. Let **A** be the linear transformation on $\mathsf{R}_2$ defined as follows:

$$\text{If } \mathbf{x} = \begin{pmatrix} x_1 \\ x_2 \end{pmatrix}, \quad \text{then } \mathbf{A}\mathbf{x} = \begin{pmatrix} x_1 - x_2 \\ x_2 - x_1 \end{pmatrix}. \tag{2.9}$$

Let

$$\mathbf{e}_1 = \begin{pmatrix} 1 \\ 1 \end{pmatrix}, \quad \mathbf{e}_2 = \begin{pmatrix} 1 \\ -1 \end{pmatrix}; \tag{2.10}$$

it is not hard to see that $N(\mathbf{A})$ consists of the set of all scalar multiples of $\mathbf{e}_1$, while $R(\mathbf{A})$ is the collection of all scalar multiples of $\mathbf{e}_2$; each is therefore a one-dimensional linear manifold in $\mathsf{R}_2$.

***Example 2.5.*** *The integral operator (2.8) on* C. Consider the linear transformation **A** on C of Example 2.3; see (2.8). Given any f in C, we may write the output g = **A**f in the form

$$g(t) = \alpha \cos t + \beta \sin t, \tag{2.11}$$

where the scalars $\alpha$ and $\beta$ are given by

$$\alpha = \int_0^\pi f(\tau) \cos \tau \, d\tau, \quad \beta = \int_0^\pi f(\tau) \sin \tau \, d\tau. \tag{2.12}$$

Thus **A** carries every function in C into a linear combination of cos t and sin t. Conversely, let $\alpha$ and $\beta$ be arbitrary scalars; if we can find an f (any f!) in C such that (2.12) both hold for the given $\alpha$, $\beta$, then we will have shown that *any* linear combination of cos t and sin t is the output of **A** for *some* f in C, and hence that $R(\mathbf{A})$ consists precisely of the set of all such linear combinations. But, as is easily verified, $f(t) = (\pi\alpha + 2\beta - 2\alpha\, t)/4$ satisfies both of (2.12), confirming our assertion about the range of **A**. Thus $R(\mathbf{A})$ is a two-dimensional linear manifold in this example. As to the null space of **A**, it is clear from (2.11), (2.12) that the image g of f under **A** is the null function if and only if f is such that both integrals in (2.12) vanish; the null space of **A** is therefore precisely the set of all such functions. [If C were rendered a real Euclidean space as in Example 1.7, we would describe $N(\mathbf{A})$ geometrically as the set of all those functions in C that are orthogonal to both cos t and sin t.] Any one of the functions f(t) = cos nt or f(t) = sin nt, n = 1, 3, 5, ..., makes both integrals in (2.12) vanish; since any finite subset of this infinite collection of functions is a linearly independent set, the dimension of the linear manifold $N(\mathbf{A})$ is infinite in this example.

A linear transformation on a vector space R is said to be *non-singular* if *both* of the following hold:

$$R(\mathbf{A}) = \mathsf{R}, \quad N(\mathbf{A}) = \{\mathbf{o}\}; \tag{2.13}$$

thus *anything* in R comes via **A** from *something* in R, and *nothing* in R that is non-null is carried by **A** to the null vector. Nonsingular transformations have simpler personalities than singular ones, but the latter are often more interesting.

***Example 2.6.*** *The linear transformation of Example 2.1.* Let **A** be the linear transformation of Example 2.1. In view of (2.4), the range of **A** is the set of all columns **y** with elements $y_1$, $y_2$ for which one can find $x_1$, $x_2$ satisfying

$$\left.\begin{aligned} a_{11}\,x_1 + a_{12}\,x_2 &= y_1, \\ a_{21}\,x_1 + a_{22}\,x_2 &= y_2; \end{aligned}\right\} \tag{2.14}$$

the null space of **A** consists of all columns **x** whose elements $x_1$, $x_2$ satisfy (2.14) with $y_1 = y_2 = 0$. An elementary consideration of the two linear equations (2.14) shows that, if $\Delta \equiv a_{11}a_{22} - a_{12}a_{21}$ is not zero, then (2.14) can be solved uniquely for $x_1$, $x_2$ *for any* $y_1$, $y_2$; in particular, necessarily $x_1 = x_2 = 0$ when $y_1 = y_2 = 0$. It then follows that, when **A** is such that $\Delta \neq 0$, one has $R(\mathbf{A}) = \mathsf{R}$, $N(\mathbf{A}) = \{\mathbf{o}\}$, so that **A** is nonsingular. When $\Delta = 0$, there are xs, not both zero, that satisfy (2.14) when $y_1 = y_2 = 0$, so that for certain columns $\mathbf{x} \neq \mathbf{o}$, one has $\mathbf{Ax} = \mathbf{o}$. Thus when $\Delta = 0$, **A** is singular, so that the transformation **A** is nonsingular if and only if $\Delta \neq 0$.

By appealing to Cramer's rule for $n \times n$ linear systems as discussed in Appendix 1, one can immediately generalize the result of Example 2.6 from $\mathsf{R}_2$ to the space $\mathsf{R}_n$ of n-rowed real columns, or for that matter to the case of the complex vector space $\mathsf{C}_n$ of n-rowed columns of complex numbers. In these vector spaces, the linear transformation analogous to **A** of Example 2.1 is associated with an $n \times n$ matrix $\underline{A}$ of real or complex numbers, as appropriate. Cramer's rule implies that the transformation is nonsingular if and only if the determinant of $\underline{A}$ is not zero. These conclusions in $\mathsf{R}_n$ and $\mathsf{C}_n$ will be of great importance in our story as it unfolds.

***Example 2.7.*** *The transformations* **A** *and* **B** *of Example 2.2.* Let **p**(t) be any *constant* polynomial in $\mathsf{P}_n$. Then if **A** is the first linear transformation described in Example 2.2, one has $\mathbf{Ap} = \mathbf{o}$, so $N(\mathbf{A}) \neq \{\mathbf{o}\}$. Thus **A** is singular. Since the only polynomial **p**(t) whose integral from zero to t vanishes *identically* is the null polynomial, the null space of the transformation **B** of Example 2.2 contains only the null vector. Moreover, given any polynomial **q** in $\mathsf{P}_n$, there is a polynomial $\mathbf{p}(t) = d/dt\,(t\,\mathbf{q}(t))$ such that $\mathbf{Bp} = \mathbf{q}$. Thus $R(\mathbf{B}) = \mathsf{P}_n$, so **B** is non-singular.

The integral operator on $\mathsf{C}$ discussed in Examples 2.3 and 2.5 clearly represents a singular linear transformation, since its null space is non-trivial. This is also the case with the linear transformation $\mathbf{A}$ on $\mathsf{R}_2$ of Example 2.4.

We now show that if $\mathbf{A}$ is a non-singular linear transformation on a vector space $\mathsf{R}$ (real or complex, finite or infinite dimensional), there is a unique linear transformation $\mathbf{B}$ on $\mathsf{R}$, also non-singular, such that $\mathbf{AB} = \mathbf{BA} = \mathbf{1}$. Conversely, if such a $\mathbf{B}$ exists, $\mathbf{A}$ is non-singular. $\mathbf{B}$ is called the inverse of $\mathbf{A}$ and is denoted by $\mathbf{A}^{-1}$. To establish this result, we proceed as follows. Suppose first that $\mathbf{A}$ is non-singular, and let $\mathbf{x}$ be any vector in $\mathsf{R}$. Since $\mathbf{A}$ is non-singular, $R(\mathbf{A}) = \mathsf{R}$, so there is a a vector $\mathbf{z}$ in $\mathsf{R}$ such that $\mathbf{Az} = \mathbf{x}$; since $N(\mathbf{A}) = \{\mathbf{o}\}$, there can be only one such $\mathbf{z}$. Clearly $\mathbf{z}$ will depend on $\mathbf{x}$, so we write $\mathbf{z} = \mathbf{z}(\mathbf{x})$. Define a transformation $\mathbf{B}$ by $\mathbf{Bx} = \mathbf{z}(\mathbf{x})$. We next show that $\mathbf{B}$ is a *linear* transformation. For any pair of vectors $\mathbf{x}$, $\mathbf{y}$ in $\mathsf{R}$, we have $\mathbf{Az}(\mathbf{x} + \mathbf{y}) = \mathbf{x} + \mathbf{y} = \mathbf{Az}(\mathbf{x}) + \mathbf{Az}(\mathbf{y})$, or $\mathbf{A}\{\mathbf{z}(\mathbf{x} + \mathbf{y}) - \mathbf{z}(\mathbf{x}) - \mathbf{z}(\mathbf{y})\} = \mathbf{o}$. Since $N(\mathbf{A}) = \{\mathbf{o}\}$, this means that $\mathbf{z}(\mathbf{x} + \mathbf{y}) - \mathbf{z}(\mathbf{x}) - \mathbf{z}(\mathbf{y}) = \mathbf{o}$, or $\mathbf{B}(\mathbf{x} + \mathbf{y}) = \mathbf{Bx} + \mathbf{By}$. Similarly, $\mathbf{B}(\alpha\mathbf{x}) = \alpha\mathbf{B}(\mathbf{x})$ for every scalar $\alpha$ and every $\mathbf{x}$ in $\mathsf{R}$, so that $\mathbf{B}$ is indeed linear. By definition of $\mathbf{B}$, we have $\mathbf{ABx} = \mathbf{x}$ for every $\mathbf{x}$ in $\mathsf{R}$, so $\mathbf{AB} = \mathbf{1}$. Furthermore, if we set $\mathbf{y} = \mathbf{BAx}$ for any $\mathbf{x}$ in $\mathsf{R}$, we have $\mathbf{ABAx} = \mathbf{Ay}$; since $\mathbf{AB} = \mathbf{1}$, this implies that $\mathbf{Ax} = \mathbf{Ay}$, and hence that $\mathbf{x} = \mathbf{y}$. Thus for every $\mathbf{x}$, $\mathbf{BAx} = \mathbf{x}$, whence $\mathbf{BA} = \mathbf{1}$ as well. We have therefore shown that if $\mathbf{A}$ is non-singular, there is a $\mathbf{B}$ with the desired properties. It is easy to show that it $\mathbf{B}$ is unique.

Conversely, suppose for a given linear transformation $\mathbf{A}$, there is a linear transformation $\mathbf{B}$ such that $\mathbf{AB} = \mathbf{BA} = \mathbf{1}$. If $\mathbf{Ax} = \mathbf{o}$, then $\mathbf{x} = \mathbf{1x} = \mathbf{BAx} = \mathbf{o}$, so that $N(\mathbf{A}) = \{\mathbf{o}\}$. Moreover, if $\mathbf{y}$ is any vector, set $\mathbf{By} = \mathbf{x}$. Then $\mathbf{Ax} = \mathbf{ABy} = \mathbf{1y} = \mathbf{y}$, showing that $R(\mathbf{A}) = \mathsf{R}$. Hence $\mathbf{A}$ is non-singular.

We shall show later in this chapter that, when $\mathsf{R}$ is finite dimensional, one has $R(\mathbf{A}) = \mathsf{R}$ if and only if $N(\mathbf{A}) = \{\mathbf{o}\}$ for any linear transformation $\mathbf{A}$, so that only one of the two qualifying conditions for non-singularity need be verified in the finite dimensional case.

It turns out that one of the most interesting and important questions about a linear transformation has to do with those features of the vector space, if any, that it leaves unchanged. For example, if one thinks of arrows in three-dimensional space, and if the linear transformation of interest is a rotation of arrow tips about a given axis through a given angle, any arrow that lies along the axis of rotation is mapped to itself, and therefore left unaffected by the action of the transformation. The concept that facilitates the study of this issue in general is that of an *invariant linear manifold*. Let $\mathbf{A}$ be a linear transformation on an arbitrary vector space $\mathsf{R}$, and let $\mathsf{M}$ be a linear manifold in $\mathsf{R}$ with the property that $\mathbf{Ax}$ is in $\mathsf{M}$ whenever $\mathbf{x}$ is in $\mathsf{M}$. We call $\mathsf{M}$ an invariant linear manifold for $\mathbf{A}$. In the example of rotation of arrows in three dimensions mentioned a moment ago, the collection of all arrows that are collinear with the axis of rotation is clearly a one-dimensional invariant linear manifold for the rotation. The set of all arrows in the plane perpendicular to the axis of rotation is a *two*-dimensional invariant linear manifold.

For any linear transformation **A** on an arbitrary vector space R, the range $R(\mathbf{A})$ and the null space $N(\mathbf{A})$ are both invariant linear manifolds of **A**.

We shall be especially interested in low-dimensional invariant linear manifolds, for reasons that will become clear soon. In particular, let us consider first the possibility that a given linear transformation **A** on an arbitrary vector space R has a *one*-dimensional invariant linear manifold, as in the case of rotations of arrows in three dimensions. If M is a one-dimensional linear manifold in R, it consists of all scalar multiples of a single non-null vector, say **x**. Such an M will be an *invariant* linear manifold for a linear transformation **A** if and only if the vector **x** has the property that

$$\mathbf{Ax} = \lambda\mathbf{x}, \quad \mathbf{x} \neq \mathbf{o}, \tag{2.15}$$

for some scalar $\lambda$. Thus to determine the one-dimensional invariant linear manifolds of a given linear transformation **A**, we must find vectors **x** such that (2.15) holds; values of $\lambda$ for which there exist vectors **x** satisfying (2.15) are called *eigenvalues* of **A**. For a given eigenvalue $\lambda$ of **A**, any vector **x** satisfying (2.15) is called an *eigenvector* of **A** corresponding to $\lambda$. The task of finding the eigenvalues and eigenvectors of **A** according to (2.15) is called the *eigenvalue problem* for **A**. One thinks of eigenvectors as descriptors of *directions* in the vector space R that are left unchanged by the action of the linear transformation **A**. Note that eigenvectors are necessarily non-null and non-unique: if **x** is an eigenvector of **A**, so is $\alpha\mathbf{x}$, where $\alpha$ is any non-zero scalar.

The qualifying phrase *for some scalar* $\lambda$ appearing after (2.15) is critical! If R is a *real* vector space, then $\lambda$ must be a real number. This innocent observation lies at the heart of the distinction between real and complex vector spaces, in a sense that will become strikingly clear as we proceed. For now, let us look at some special examples to get some idea of what the possibilities are.

Suppose the null space of a linear transformation **A** on an arbitrary vector space R is non-trivial in the sense that there is a vector $\mathbf{x} \neq \mathbf{o}$ in R such that $\mathbf{Ax} = \mathbf{o}$; then clearly **A** is singular, and **x** is an eigenvector of **A** with eigenvalue $\lambda = 0$. Conversely, if $\lambda = 0$ is an eigenvalue of **A**, then the null space of **A** is non-trivial, so **A** must be singular.

As we shall see, eigenvalues and eigenvectors together are like the "genetic code" of a linear transformation: they go a long way toward determining the personality and behavior of **A**.

***Example 2.8.*** *The linear transformation (2.4) on* $R_2$. Let **A** be the linear transformation on the *real* vector space $R_2$ introduced in Example 2.1.

Clearly a column **x** with entries $x_1$ and $x_2$ will be an eigenvector of **A** if it is non-null and such that

$$\left.\begin{aligned}(a_{11} - \lambda)x_1 + a_{12}x_2 = 0,\\ a_{21}x_1 + (a_{22} - \lambda)\,x_2 = 0,\end{aligned}\right\} \tag{2.16}$$

for some scalar $\lambda$. Since the equations for $x_1$, $x_2$ in (2.16) comprise a linear homogeneous system, and since $x_1^2 + x_2^2 \neq 0$, an eigenvalue $\lambda$ must be such that the determinant of the system vanishes: thus

$$\lambda^2 - (a_{11} + a_{22})\,\lambda + a_{11}a_{22} - a_{12}a_{21} = 0. \tag{2.17}$$

One possibility is that (2.17) has no real roots $\lambda$. If this occurs, then **A** has *no* eigenvalues or eigenvectors; complex roots $\lambda$ of (2.17) do not qualify as scalars for the space $\mathsf{R}_2$ and hence will not serve as eigenvalues in (2.16). The condition that the roots $\lambda$ of (2.17) be real is that the relevant discriminant be non-negative:

$$(a_{11} - a_{22})^2 + 4\,a_{12}a_{21} \geq 0. \tag{2.18}$$

The detailed possibilities accompanying (2.17), (2.18) are worked out in Problem 2.6; we summarize the results here. Suppose first that *strict* inequality holds in (2.18). Then (2.17) has two distinct real roots $\lambda_1$, $\lambda_2$, and the system (2.16) has solutions for each of these $\lambda$s that provide corresponding *linearly independent* eigenvectors $\mathbf{x} = \mathbf{e}_1$, $\mathbf{x} = \mathbf{e}_2$ of **A** satisfying (2.15). But now suppose that *equality* holds in (2.18). Then the two real roots $\lambda$ of (2.17) coincide, and a mathematical nightmare results! As shown by the analysis necessary to work out Problem 2.6, there *may* be two linearly independent eigenvectors anyway, but then again there may not.

***Example 2.9.*** *The linear transformations of Example 2.2.* Consider the vector space $\mathsf{P}_n$ of real polynomials of degree at most $n - 1$, and let **A** and **B** be the linear transformations of Example 2.2. A polynomial **p** and an real number $\lambda$ are respectively an eigenvector and an eigenvalue of **A** if

$$\mathbf{A}\mathbf{p} = \lambda\mathbf{p}; \tag{2.19}$$

from the definition (2.5) of **A**, it is easily shown that $\lambda = 0$ is the only eigenvalue of **A**, and the corresponding eigenvector is the constant polynomial. For the transformation **B**, the story is a bit more complicated: from the definition (2.6) of **B**, it follows that the eigenvalue problem for **B** is equivalent to the integral equation

$$\lambda t\,\mathbf{p}(t) = \int_0^t \mathbf{p}(t)\,dt, \quad t \geq 0, \tag{2.20}$$

for a polynomial $\mathbf{p}(t)$ of degree $n - 1$. Suppose $\mathbf{p}(t)$ satisfies (2.20). Then if $\mathbf{p}(t) = \alpha_0 + \cdots + \alpha_N t^N$, where $N = n - 1$, one finds from (2.20) that necessarily $\alpha_k = (k + 1)\lambda\ \alpha_k$, for $k = 0, 1, \ldots, N$. It follows that the $j^{th}$ eigenvalue $\lambda_j$ is given by $\lambda_j = (1/(j + 1)$, with the corresponding eigenvector $\mathbf{p}(t)$ proportional to the special polynomial $\mathbf{p}_j(t) = t^j$, $j = 0, \ldots, N$. (Problem 2.8 addresses an interesting question: what are the values, if any, of the real number $\lambda$ for which non-null solutions of the integral equation (2.20) exist that, while not polynomials, are nevertheless continuous functions for $t \geq 0$?)

***Example 2.10.*** *The integral operator on* C *of Example 2.3.* Clearly a continuous function f on the interval $[0, \pi]$ is an eigenvector—or perhaps more suggestively an eigen*function*—of the transformation $\mathbf{A}$ of (2.8) if there is a real number $\lambda$, the corresponding eigenvalue, such that

$$\lambda\, f(t) = \int_0^{\pi} \cos(t - \tau)\, f(\tau)\, d\tau, \quad 0 \leq t \leq \pi. \tag{2.21}$$

We need to find the non-trivial solutions f of (2.21). We first note that $\lambda = 0$ is certainly an eigenvalue, with any f in the null space of $\mathbf{A}$ serving as a corresponding eigenfunction; see Example 2.5. Thus there are infinitely many eigenfunctions corresponding to the eigenvalue $\lambda = 0$. Next suppose that $\lambda \neq 0$; then (2.21) says that

$$f(t) = \frac{\alpha}{\lambda} \cos t + \frac{\beta}{\lambda} \sin t, \tag{2.22}$$

with $\alpha$, $\beta$ given by (2.12). Thus f is necessarily a linear combination of cos t and sin t. Substituting f as given by (2.22) into both sides of (2.21) shows that $\alpha$, $\beta$ and $\lambda$ must satisfy

$$\alpha\,(\lambda - \pi/2) = 0, \quad \beta\,(\lambda - \pi/2) = 0. \tag{2.23}$$

To assure that f is non-trivial, we must have $\lambda = \pi/2$, in which case both $\alpha$ and $\beta$ are arbitrary. Thus $\lambda = \pi/2$ is an eigenvalue, and the corresponding eigenfunction is an arbitrary linear combination of cos t and sin t.

We now comment briefly on *two*-dimensional invariant linear manifolds; these will prove to be of special interest shortly. Let M be a two-dimensional linear manifold that is invariant for the linear transformation $\mathbf{A}$. Then there must be linearly independent vectors $\mathbf{x}$ and $\mathbf{y}$, forming a basis for M, and scalars $\alpha$, $\beta$, $\gamma$, and $\delta$ such that

$$\mathbf{A}\mathbf{x} = \alpha\,\mathbf{x} + \beta\,\mathbf{y}, \quad \mathbf{A}\mathbf{y} = \gamma\,\mathbf{x} + \delta\,\mathbf{y}. \tag{2.24}$$

If both $\beta$ and $\gamma$ vanish, then $\mathbf{x}$ and $\mathbf{y}$ are eigenvectors of $\mathbf{A}$, and $\mathsf{M}$ contains two one-dimensional invariant linear manifolds; otherwise $\mathsf{M}$ is *irreducible*, in the sense that it contains *no* one-dimensional invariant linear manifold for $\mathbf{A}$; see Problem 2.14. The role played by these more complicated two-dimensional creatures $\mathsf{M}$ will become clear soon.

## Linear Transformations on Finite-Dimensional Spaces

The remainder of this chapter will be concerned with finite-dimensional vector spaces. Our objective now is to exploit the assumption of finite dimensionality to probe the nature of linear transformations in this setting. As yet we make no use of the metric properties of Euclidean spaces; the theory of linear transformations in *these* rich environments is precisely the theory of Cartesian tensors; we defer their discussion to the next chapter.

We begin by studying the representation of linear transformations on finite-dimensional real or complex vector spaces. This study not only leads to results that are useful for calculation, but also provides tools that permit us to uncover more about the nature of non-singular linear transformations. We then make further use of these tools to illuminate the situation concerning low-dimensional invariant linear manifolds of a linear transformation, making clear in the process the essential difference between real and complex finite-dimensional vector spaces.

Let $\mathsf{R}$ be a real or complex linear vector space of finite dimension n, and let $\mathbf{e}_1, ..., \mathbf{e}_n$ be a basis for $\mathsf{R}$. If $\mathbf{A}$ is any linear transformation, $\mathbf{Ae}_j$ is in $\mathsf{R}$ for each j, so there are $n^2$ scalars $a_{kj}$, k, j = 1, ..., n, such that

$$\mathbf{Ae}_j = a_{kj}\,\mathbf{e}_k; \tag{2.25}$$

the range and summation conventions are in force. The numbers $a_{kj}$ are called the *components of* $\mathbf{A}$ *in the basis* $\mathbf{e}$; to emphasize the dependence of these numbers on the choice of basis, we shall sometimes denote them by $a^{\mathbf{e}}_{kj}$. Since this is rather heavy notational baggage, we shall omit the superscript $\mathbf{e}$ whenever it is not dangerous to do so. When *two* or more bases figure in an argument or a calculation, there *is* danger, and we will use the more cumbersome designation in such situations.

It is also useful to introduce the *matrix of components of* $\mathbf{A}$ *in the basis* $\mathbf{e}$, denoted by $\underline{A}$ (or by $\underline{A}^{\mathbf{e}}$ if dependence on $\mathbf{e}$ needs emphasis), by setting

$$\underline{A} = \begin{pmatrix} a_{11} & a_{12} & \cdots & a_{1n} \\ a_{21} & a_{22} & \cdots & a_{2n} \\ \cdots & \cdots & \cdots & \cdots \\ a_{n1} & a_{n2} & \cdots & a_{nn} \end{pmatrix} \tag{2.26}$$

note that each a—upper *and* lower case—in (2.26) can be decorated with superscript **e**s if basis-dependence is to be exhibited.

Suppose vectors **x** and **y** are related by $\mathbf{y} = \mathbf{A}\mathbf{x}$; let $\xi_i$ and $\eta_i$ be the components in **e** of **x** and **y**, respectively. Then the assertion $\mathbf{y} = \mathbf{A}\mathbf{x}$ can be written in the equivalent form

$$\eta_i \, \mathbf{e}_i = \mathbf{A}(\xi_j \, \mathbf{e}_j) = a_{ij} \, \xi_j \, \mathbf{e}_i, \tag{2.27}$$

or, since the **e**s are linearly independent,

$$\eta_i = a_{ij} \, \xi_j. \tag{2.28}$$

Thus $\mathbf{y} = \mathbf{A}\mathbf{x}$ implies the n scalar relations (2.28) among the components of **x**, **y**, and **A**; it is easy to see that, conversely, the relations (2.28) imply (2.27), so that the basis-free vector statement $\mathbf{y} = \mathbf{A}\mathbf{x}$ is equivalent to the relations (2.28) among basis-dependent components. We think of the scalar relations (2.28) as arising from the *resolution into components in* **e** of the vector relation $\mathbf{y} = \mathbf{A}\mathbf{x}$.

Suppose that **A** and **B** are linear transformations on $\mathsf{R}$ with respective matrices of components $\underline{A}$ and $\underline{B}$ in a given basis **e**. The following dull questions need to be answered: (a) If $\mathbf{C} = \alpha \, \mathbf{A}$ for some scalar $\alpha$, what is $\underline{C}$? (b) If $\mathbf{C} = \mathbf{A} + \mathbf{B}$, what is $\underline{C}$? (c) If $\mathbf{C} = \mathbf{AB}$, what is $\underline{C}$? (d) If **A** is non-singular and $\mathbf{C} = \mathbf{A}^{-1}$, what is $\underline{C}$? The first two questions are easily shown to have the predictable answers: for (a), $\underline{C} = \alpha \, \underline{A}$, and for (b) $\underline{C} = \underline{A} + \underline{B}$; it is assumed that the reader's familiarity with matrices (as distinguished from linear transformations) makes clear the meaning of the last two assertions. One finds the answer to question (c) by resolving the vector relation $\mathbf{Cx} = \mathbf{ABx}$ into components in the basis **e**, leading to $c_{ij} \, \xi_j = a_{ik} b_{kj} \, \xi_j$—remember the summation convention! Since the $\xi$'s are arbitrary, one concludes that $c_{ij} = a_{ik} \, b_{kj}$. This represents the usual rule for matrix multiplication; indeed, it explains why that rule is the way it is. (It is not the *only* useful way to invent matrix multiplication; for a discussion of other ways, see the book by Richard Bellman [2.1] listed among the references at the end of this chapter. The *usual* way, though, is the only one of interest to us, because of our focus on linear transformations.)

Unfortunately, we cannot escape question (d)! As we showed in the preceding chapter, if **A** is non-singular, then it has an inverse $\mathbf{A}^{-1}$; let $\mathbf{C} = \mathbf{A}^{-1}$. What are the components $c_{ij}$ of **C** in a given basis $\mathbf{e}_1, \ldots, \mathbf{e}_n$? Since **A** is non-singular, for any **y** in $\mathsf{R}$, there is an **x** in $\mathsf{R}$ such that $\mathbf{y} = \mathbf{A}\mathbf{x}$. But we explained above that, in a finite dimensional space, this equation is equivalent to (2.27), where $\eta_i$ and $\xi_i$ are the respective components of **y** and **x** in the basis **e**. Moreover, $\mathbf{x} = \mathbf{C}\mathbf{y}$, so

$$\xi_i = c_{ij} \, \eta_j. \tag{2.29}$$

Let $\underline{A}$ and $\underline{C}$ be the respective matrices of components of $\mathbf{A}$ and $\mathbf{C} = \mathbf{A}^{-1}$ in the basis $\mathbf{e}$. Since $\mathbf{AC} = \mathbf{1}$, the answer to question (c) implies that $\underline{AC} = \underline{1}$, where $\underline{1}$ is the identity matrix. Among the properties of determinants listed in Appendix 1, there is one that says det $\underline{AC}$ = det $\underline{A}$ det $\underline{C}$; since det $\underline{1} = 1$, it follows that in the present circumstance, det $\underline{A}$ det $\underline{C} = 1$, whence it must be so that neither det $\underline{A}$ nor det $\underline{C}$ can vanish. Thus the determinant of the linear system (2.28) is not zero, so that by Cramer's rule as set out in Appendix 1, this system has a unique solution given by

$$\xi_i = \frac{1}{\det \underline{A}} A_{ji}\, \eta_j, \tag{2.30}$$

where $A_{ji}$ is the cofactor of the element $a_{ji}$ in the matrix $\underline{A}$. Comparing (2.30) and (2.29) and bearing in mind that the $\eta_i$s are arbitrary, we conclude that the components $c_{ij}$ in $\mathbf{e}$ of the inverse of $\mathbf{A}$ are given by

$$c_{ij} = \frac{1}{\det \underline{A}} A_{ji}. \tag{2.31}$$

Thus the blue-collar questions (a)–(d), indispensable for calculation, have been answered.

There is an important parenthetical remark to be made here: the detailed calculations made in the forgoing argument appeared to depend on the choice of basis $\mathbf{e}$. The fact that det $\underline{A}^{\mathbf{e}} \neq 0$ was crucial in this reasoning. Is it possible that this determinant, though different from zero in the basis $\mathbf{e}$, might vanish in some other basis? We show a few paragraphs in the future that this cannot occur; in fact, det $\underline{A}^{\mathbf{e}}$ has the same value in all bases, so that det $\mathbf{A}$ is a scalar whose value is determined by the linear transformation $\mathbf{A}$ itself. We take this for granted for the next few paragraphs for reasons of simplicity of presentation.

We return now to the subject of non-singular linear transformations on finite dimensional spaces, establishing first a result that was promised earlier in this chapter. We show that, for a given linear transformation $\mathbf{A}$, one has $N(\mathbf{A}) = \{\mathbf{o}\}$ if and only if $R(\mathbf{A}) = \mathsf{R}$, making it necessary to check only one condition, rather than two, in order to verify the non-singularity of $\mathbf{A}$. As one might expect, the argument to be given below depends critically on the existence of a basis, and hence for us is applicable only to finite dimensional spaces. Accordingly, assume that $\mathsf{R}$ is finite dimensional, and suppose first that $R(\mathbf{A}) = \mathsf{R}$. Let $\mathbf{e}_1, \ldots, \mathbf{e}_n$ be a basis for $\mathsf{R}$, and observe that there are vectors $\mathbf{g}_i$ such that $\mathbf{A}\mathbf{g}_i = \mathbf{e}_i$. The $\mathbf{g}_i$s must be linearly independent, because if there were scalars $\alpha_1, \ldots, \alpha_n$ such that $\alpha_i\, \mathbf{g}_i = \mathbf{o}$ (don't forget to sum), then one would have $\mathbf{A}(\alpha_i \mathbf{g}_i) = \mathbf{o}$, or $\alpha_i \mathbf{e}_i = \mathbf{o}$, whence all the $\alpha_i$s would necessarily vanish, by the independence of the $\mathbf{e}_i$s. Thus the $\mathbf{g}_i$s form a basis for

R, too. Suppose now that $\mathbf{x}$ is such that $\mathbf{Ax} = \mathbf{o}$. Let $\mathbf{x} = \zeta_i \mathbf{g}_i$; then $\zeta_i \mathbf{A}\mathbf{g}_i = \zeta_i \mathbf{e}_i = \mathbf{o}$, so all the $\zeta_i$s must coincide with the null scalar. Hence $N(\mathbf{A}) = \{\mathbf{o}\}$. Having shown that $R(\mathbf{A}) = \mathsf{R}$ implies $N(\mathbf{A}) = \{\mathbf{o}\}$, we need only show that if $N(\mathbf{A}) = \{\mathbf{o}\}$, then also $R(\mathbf{A}) = \mathsf{R}$ to establish the result. Indeed, suppose that $N(\mathbf{A}) = \{\mathbf{o}\}$, and let $\mathbf{f}_i = \mathbf{A}\mathbf{e}_i$. Let $\mathbf{y}$ be any vector in R, and represent $\mathbf{y}$ as $\mathbf{y} = \eta_i \mathbf{f}_i$. Put $\mathbf{x} = \eta_i \mathbf{e}_i$; then clearly $\mathbf{Ax} = \eta_i \mathbf{A}\mathbf{e}_i = \eta_i \mathbf{f}_i = \mathbf{y}$. Thus for any $\mathbf{y}$ in R, there is an $\mathbf{x}$ in R such that $\mathbf{Ax} = \mathbf{y}$, whence $R(\mathbf{A}) = \mathsf{R}$, and the matter is resolved.

It is important to point out that we have demonstrated the following result in the discussion given above: if $\mathbf{e}_1, \ldots, \mathbf{e}_n$ is a basis for R and if $\mathbf{A}$ is a *non-singular* linear transformation, then the vectors $\mathbf{f}_i = \mathbf{A}\mathbf{e}_i$ *also* comprise a basis for R. The converse is also true; if the $\mathbf{e}_i$s and $\mathbf{f}_i$s are both bases and if $\mathbf{R}$ is a linear transformation for which $\mathbf{R}\mathbf{e}_i = \mathbf{f}_i$, then $\mathbf{R}$ is non-singular; this is the subject of Problem 2.9.

A simple argument shows that a linear transformation $\mathbf{A}$ on a finite dimensional vector space is non-singular if and only if $\det \mathbf{A} \neq 0$.

We turn next to the *change-of-basis formula* for the components of a linear transformation on a finite-dimensional vector space. In Chapter 1, we derived such a formula for the components of a vector; see (1.9). We produce here the precise counterpart of (1.9) for the components of a linear transformation. Let $\mathbf{A}$ be a linear transformation on R, and let $\mathbf{e}_1, \ldots, \mathbf{e}_n$ and $\mathbf{f}_1, \ldots, \mathbf{f}_n$ be two bases for R. Let $a^{\mathbf{e}}_{ij}$ and $a^{\mathbf{f}}_{ij}$ be the components of $\mathbf{A}$ in $\mathbf{e}$ and $\mathbf{f}$, respectively. What is the relation between these two sets of components? To help answer this question, we invent a linear transformation $\mathbf{R}$ on R as follows: if $\mathbf{x} = \xi_i \mathbf{e}_i$, let $\mathbf{Rx} = \xi_i \mathbf{f}_i$; in particular, $\mathbf{R}\mathbf{e}_i = \mathbf{f}_i$. (Is $\mathbf{R}$ *really* linear?) By definition,

$$\mathbf{A}\mathbf{f}_i = a^{\mathbf{f}}_{ki}\mathbf{f}_k \quad \text{and} \quad \mathbf{A}\mathbf{e}_i = a^{\mathbf{e}}_{ki}\,\mathbf{e}_k. \tag{2.32}$$

But on the one hand,

$$\mathbf{A}\mathbf{f}_i = a^{\mathbf{f}}_{ki}\mathbf{f}_k = a^{\mathbf{f}}_{ki}\,\mathbf{R}\mathbf{e}_k = a^{\mathbf{f}}_{ki}\,r^{\mathbf{e}}_{mk}\mathbf{e}_m, \tag{2.33}$$

where $r^{\mathbf{e}}_{mk}$ are the components of $\mathbf{R}$ in $\mathbf{e}$. On the other hand,

$$\mathbf{A}\mathbf{f}_i = \mathbf{AR}\,\mathbf{e}_i = \mathbf{A}r^{\mathbf{e}}_{ji}\mathbf{e}_j = r^{\mathbf{e}}_{ji}\,\mathbf{A}\mathbf{e}_j = r^{\mathbf{e}}_{ji}\,a^{\mathbf{e}}_{mj}\,\mathbf{e}_m. \tag{2.34}$$

Comparing (2.33) and (2.34) yields

$$a^{\mathbf{f}}_{ki}\,r^{\mathbf{e}}_{mk} = r^{\mathbf{e}}_{ji}\,a^{\mathbf{e}}_{mj}. \tag{2.35}$$

In terms of the various matrices of components of the tensors involved, we may write (2.35) in the equivalent matrix form

$$\underline{R}^{\mathbf{e}}\,\underline{A}^{\mathbf{f}} = \underline{A}^{\mathbf{e}}\,\underline{R}^{\mathbf{e}}; \tag{2.36}$$

By the remark in the preceding paragraph, the determinant of the matrix $\underline{R}^{e}$ cannot vanish so that $\underline{R}^{e}$ has an inverse. The matrix equation (2.36) can therefore be rewritten in the form

$$\underline{A}^{f} = (\underline{R}^{e})^{-1}\, \underline{A}^{e}\, \underline{R}^{e}. \tag{2.37}$$

The relation (2.37) yields the matrix of components of **A** in the basis **f** in terms of the matrix of **A** in the basis **e**, with the matrix in **e** of the "basis conversion" tensor **R** playing the role of intermediary. We call (2.37) the change-of-basis formula for the components of the linear transformation **A**.

Among the properties of determinants listed in Appendix 1, there are two that are useful here: one says the determinant of the product of two matrices is the product of their separate determinants, while the other asserts that the determinant of the inverse of a non-singular matrix is the reciprocal of the value of the determinant of the original matrix. By making use of these facts in (2.37), we can assert that

$$\det \underline{A}^{f} = \det \underline{A}^{e}, \tag{2.38}$$

so that the determinant of the matrix of components of a linear transformation is the same in *all* bases, confirming a result taken for granted earlier. The determinant may thus be viewed as a scalar-valued function of the components of a linear transformation whose value is invariant under changes of basis.

Another example of such a *scalar invariant* associated with a linear transformation **A** is its *trace*: the trace of **A**, written Tr **A**, is the sum of the diagonal elements of its matrix of components in any basis; thus Tr $\mathbf{A} = a^{e}_{ii}$. According to the change-of-basis formula (2.37), we have $a^{f}_{ii} = r^{-1}_{ij}\, a^{e}_{jk}\, r_{ki}$, where $r_{ij}$ and $r^{-1}_{ij}$ are the respective components in **e** of **R** and $\mathbf{R}^{-1}$. According to Appendix 1, these quantities are such that $r_{ki}\, r^{-1}_{ij} = \delta_{kj}$, so that indeed $a^{f}_{ii} = a^{e}_{ii}$.

One way to specify a particular linear transformation **A** on a finite-dimensional vector space is to pick a basis **e** and prescribe arbitrarily the values of the components $a^{e}_{ij}$ of **A**, and therefore the matrix $\underline{A}^{e}$, in this basis. The values of the components $a^{f}_{ij}$, and therefore the matrix $\underline{A}^{f}$, are then determined in any other basis **f** by the change-of-basis formulas (2.35) and (2.36), respectively.

We now exploit the availability of a basis in any finite-dimensional vector space to establish some interesting results about invariant linear manifolds. Let **A** be a linear transformation on an arbitrary finite-dimensional vector space $\mathsf{R}$, and suppose that $\mathsf{M}$ is a one-dimensional invariant linear manifold for **A**. Then according to (2.15), there is a vector **x** in $\mathsf{M}$ such that

$$\mathbf{Ax} = \lambda\, \mathbf{x}, \quad \mathbf{x} \neq \mathbf{o}. \tag{2.39}$$

If $\mathbf{e}_1, ..., \mathbf{e}_n$ is a basis for R, we may represent $\mathbf{x}$ in (2.39) in the form $\mathbf{x} = \xi_k \mathbf{e}_k$, thus reducing the problem (2.39) for $\mathbf{x}$ to the problem of finding those values of $\lambda$, if any, for which the system of linear algebraic equations

$$a_{jk} \xi_k = \lambda \xi_j \tag{2.40}$$

has a solution in which not all the $\xi_j$'s are zero. According to Cramer's rule as described in Appendix 1, there is a non-trivial solution $\xi_1, ..., \xi_n$ of (2.40) if and only if $\lambda$ satisfies

$$P_{\mathbf{A}}(\lambda) \equiv \det(\underline{A}^e - \lambda \underline{1}^e) = \det(\mathbf{A} - \lambda \mathbf{1}) = 0, \tag{2.41}$$

where $\underline{1}^e$ is the n × n identity matrix. Elementary properties of determinants reveal that, as a function of $\lambda$, $P_{\mathbf{A}}(\lambda)$ is a polynomial of degree n. In view of our earlier discussion, $P_{\mathbf{A}}$ is a function of $\lambda$ that depends on $\mathbf{A}$, but not on the basis $\mathbf{e}$. We call $P_{\mathbf{A}}(\lambda)$ the *characteristic polynomial* of $\mathbf{A}$; its coefficients are scalars. The linear transformation $\mathbf{A}$ has a one-dimensional invariant linear manifold if and only if $P_{\mathbf{A}}(\lambda)$ has a zero $\lambda$ *that belongs to the set of scalars associated with the underlying vector space.*

We now come to an interesting fork in the road: on the one hand, according to the fundamental theorem of algebra, every polynomial of degree $n \geq 1$ with complex coefficients has at least one zero (in general complex), so we can immediately assert the following proposition:

***Proposition 2.1.*** Every linear transformation on a finite-dimensional complex vector space has at least one one-dimensional invariant linear manifold.

On the other hand, if the vector space is *real*, the story is more complicated; because polynomials with real coefficients need not have real zeros, the relevant proposition now looks like this:

***Proposition 2.2.*** Every linear transformation on a finite-dimensional real vector space has *either* a one-dimensional *or* a two-dimensional invariant linear manifold.

We now prove the second of these propositions; in doing so, we will discover the meaning of a *complex* zero of $P_{\mathbf{A}}(\lambda)$ when the vector space is real. If $P_{\mathbf{A}}(\lambda)$ has a real zero $\lambda$, then by Cramer's rule, the system (2.40) has a non-trivial solution $\xi_1, ..., \xi_n$ for this $\lambda$ in which the $\xi_k$s are all real and do not all vanish. Setting $\mathbf{x} = \xi_k \mathbf{e}_k$ then provides an eigenvector of $\mathbf{A}$, which in turn generates a one-dimensional invariant linear manifold for $\mathbf{A}$. On the other hand, suppose that $P_A(\lambda)$ has a *complex* zero $\lambda = \mu + i\nu$, where $i = (-1)^{1/2}$ is the imaginary unit, and $\mu$, $\nu$ are real, with $\nu \neq 0$. Then Cramer's rule again guarantees that (2.40), viewed momentarily as a *complex* linear system divorced from the issue of finding eigenvectors of **A**, has a non-trivial solution, but now this solution will in general be complex: thus $\xi_k = \alpha_k + i\,\beta_k$, where the $\alpha_k$s and $\beta_k$s are real. Set $\lambda = \mu + i\,\nu$ and $\xi_k = \alpha_k + i\,\beta_k$ in (2.40) and equate first the *real* parts, then the *imaginary* parts, of both sides of the resulting equations, bearing in mind that the quantities $a_{jk}$ are real. The result is a *real* system of 2n equations for the 2n unknowns $\alpha_1, ..., \alpha_n, \beta_1, ..., \beta_n$ containing $\mu$ and $\nu$ as parameters:

$$a_{jk}\,\alpha_k = \mu\,\alpha_j - \nu\,\beta_j, \quad a_{jk}\,\beta_k = \nu\,\alpha_j + \mu\,\beta_j. \tag{2.42}$$

After this "closet calculation" with complex numbers, let us return to the reality of our vector space R and construct two vectors **x** and **y** as follows:

$$\mathbf{x} = \alpha_k\,\mathbf{e}_k, \quad \mathbf{y} = \beta_k\,\mathbf{e}_k. \tag{2.43}$$

We note in passing that **x** and **y** cannot both be null vectors. Now $\mathbf{Ax} = \alpha_k\,\mathbf{Ae}_k = \alpha_k\,a_{jk}\,\mathbf{e}_j$, so making use of $(2.42)_1$ yields $\mathbf{Ax} = \mu\,\alpha_j\,\mathbf{e}_j - \nu\,\beta_j\,\mathbf{e}_j = \mu\,\mathbf{x} - \nu\,\mathbf{y}$; similarly $\mathbf{Ay} = \nu\,\mathbf{x} + \mu\,\mathbf{y}$. Thus we have constructed two vectors **x** and **y** that together satisfy

$$\mathbf{Ax} = \mu\,\mathbf{x} - \nu\,\mathbf{y}, \quad \mathbf{Ay} = \nu\,\mathbf{x} + \mu\,\mathbf{y}, \tag{2.44}$$

which is a special case of the system (2.24). Therefore if we can show that **x** and **y** are linearly independent, we may conclude that **x** and **y** represent generators of the *two*-dimensional invariant linear manifold M = span (**x**, **y**) for **A**. As to the question of linear independence, suppose there were scalars, say $\kappa$ and $\omega$, such that

$$\kappa\,\mathbf{x} + \omega\,\mathbf{y} = \mathbf{o}. \tag{2.45}$$

Then $\kappa\,\mathbf{Ax} + \omega\,\mathbf{Ay} = \mathbf{o}$ as well, so by (2.44),

$$(\kappa\mu + \omega\nu)\,\mathbf{x} + (-\kappa\nu + \omega\mu)\,\mathbf{y} = \mathbf{o}. \tag{2.46}$$

From (2.45) and (2.46) it follows that

$$\nu(\kappa^2 + \omega^2)\,\mathbf{x} = -\nu(\kappa^2 + \omega^2)\,\mathbf{y} = \mathbf{o}. \tag{2.47}$$

Since $\nu \neq 0$ and not both **x** and **y** are null, one must have $\kappa^2 + \omega^2 = 0$, so that $\kappa = \omega = 0$, and **x** and **y** are indeed linearly independent. Thus every linear transformation of a real vector space has either a one- or a two-dimensional invariant linear manifold, as asserted in the proposition.

We have shown that, when the vector space R is real, a complex zero $\lambda = \mu + i\nu$ of the characteristic polynomial of **A** corresponds to a two-dimensional invariant linear manifold M for **A**. Since $P_{\mathbf{A}}(\lambda)$ has real coefficients when R is real, the complex zeros of $P_{\mathbf{A}}(\lambda)$, if any, occur in conjugate pairs. Thus if $\lambda = \mu + i\nu$ is a zero, so is $\lambda = \mu - i\nu$. It is a simple matter to show that the two-dimensional invariant linear manifold for **A** determined by $\lambda = \mu - i\nu$ is identical to the manifold determined by $\lambda = \mu + i\nu$. Finally, it should be noted that in the proof above, we used the fact that $\nu = \text{Im}\,\lambda$ was not zero. If $\nu$ *were* zero, (2.44) would assert that **x** and **y** are eigenvectors of **A** corresponding to the eigenvalue $\mu$, and therefore are generators of two *one*-dimensional invariant linear manifolds for **A**.

The argument used above to prove the second proposition raises a question: for a linear transformation **A** on a real vector space R, could there be irreducible two-dimensional invariant linear manifolds for **A** that do *not* arise from complex zeros of $P_{\mathbf{A}}(\lambda)$? According to Problems 2.12 and 2.13, the answer to this question is no.

***Example 2.11.*** *Special linear transformations on* $R_3$ *and* $C_3$. Consider first the real vector space $R_3$ of columns **x** of three real numbers, and let **A** be the linear transformation defined by

$$\mathbf{Ax} = \begin{pmatrix} 1 & 0 & 0 \\ 0 & \cos\varphi & -\sin\varphi \\ 0 & \sin\varphi & \cos\varphi \end{pmatrix} \begin{pmatrix} x_1 \\ x_2 \\ x_3 \end{pmatrix}, \tag{2.48}$$

where $\varphi$ is a real number that is not an integral multiple of $\pi$. If we identify $R_3$ with ordinary three-dimensional point space, **A** is a rotation about the 1-axis through an angle $\varphi$. The characteristic polynomial $P_{\mathbf{A}}(\lambda) = \det(\mathbf{A} - \lambda\mathbf{1})$ of **A** is readily found to be

$$P_{\mathbf{A}}(\lambda) = (1 - \lambda)(\lambda^2 - 2\lambda\cos\varphi + 1). \tag{2.49}$$

Since the second factor on the right in (2.49) does not vanish for any real $\lambda$, the only real zero of $P_{\mathbf{A}}(\lambda)$—and therefore the only eigenvalue of **A**—is $\lambda = 1$; the corresponding one-dimensional invariant linear manifold coincides with the axis of rotation, which is a physically natural result. The two-dimensional invariant linear manifold arising from either of the complex conjugate zeros of the second factor in (2.49) is a plane *perpendicular* to the axis of rotation. Since a vector in this plane remains in the plane after rotation, this too is physically understandable.

Now suppose that we replace $R_3$ in Example 2.11 by the space $C_3$ of columns of three *complex* numbers $x_1, x_2, x_3$, the scalars being complex as well. Even with $\varphi$ real, we may reconsider **A** of (2.48) as a linear transformation on $C_3$. The characteristic polynomial for **A** continues to be that in (2.49), but now we must view the zeros differently. These are $\lambda = 1$, $\lambda = e^{i\varphi}$ and $\lambda = e^{-i\varphi}$. Each of these zeros is now an eigenvalue for **A**, and each of the corresponding eigenvectors generates a *one*-dimensional invariant linear manifold for **A**.

The distinction between real and complex vector spaces just described is often not emphasized in treatments of our subject intended primarily for mathematicians, since such treatments frequently deal only with the complex case; see, for example, the book on finite dimensional vector spaces by Halmos [2.4] cited at the end of this chapter. On the other hand, Section 16 of Gel'fand's *Lectures on Linear Algebra* [2.3], also listed among the references for this chapter, is explicitly devoted to linear transformations on real Euclidean spaces.

Let **A** be a linear transformation on a finite dimensional vector space, and let $\lambda$ be an eigenvalue of **A**. Let m be the order of $\lambda$ as a zero of the characteristic polynomial of **A**; m is called the *algebraic multiplicity* of $\lambda$. Suppose that there are exactly k linearly independent eigenvectors $\mathbf{x}_1, ..., \mathbf{x}_k$ of **A** corresponding to $\lambda$: $\mathbf{A}\mathbf{x}_j = \lambda\mathbf{x}_j$, $j = 1, ..., k$. Let $\mathsf{M} = \text{span}\,(\mathbf{x}_1, ... \mathbf{x}_k)$, so that $\mathsf{M}$ is a k-dimensional invariant linear manifold for **A**. One speaks of k as the *geometric multiplicity* of $\lambda$. The following example shows that these two notions of the multiplicity of an eigenvalue need not coincide.

***Example 2.12.*** Let $\mathsf{R}$ be a two-dimensional real vector space, and let **A** be a linear transformation on $\mathsf{R}$ whose matrix of components in a certain basis $\mathbf{e}_1, \mathbf{e}_2$ is given by

$$\underline{A} = \begin{pmatrix} 1 & 1 \\ 0 & 1 \end{pmatrix}. \tag{2.50}$$

The characteristic polynomial of **A** is given by $P_{\mathbf{A}}(\lambda) = (1 - \lambda)^2$, so that $\lambda = 1$ is an eigenvalue of **A** of algebraic multiplicity $m = 2$. Let **x** be an eigenvector of **A** corresponding to $\lambda = 1$. It is easily shown that **x** is necessarily a scalar multiple of $\mathbf{e}_1$, so that $\mathsf{M}$ is one-dimensional in this case, and the geometric multiplicity of $\lambda = 1$ is therefore $k = 1 \neq m$.

Let $\mathsf{R}$ be a vector space, real or complex, of finite dimension n, and let **A** be a linear transformation on $\mathsf{R}$ with n *distinct* eigenvalues $\lambda_1, ..., \lambda_n$ and n corresponding eigenvectors $\mathbf{e}_1, ..., \mathbf{e}_n$. One can show (Problem 2.17) that the distinctness of the eigenvalues implies the linear independence of the eigenvectors, so that the $\mathbf{e}_i$s form a basis for $\mathsf{R}$. Since $\mathbf{A}\mathbf{e}_i = \lambda_i\mathbf{e}_i$ (no sum on i), the matrix $\underline{A}^e$ of **A** in the basis **e** is diagonal, with the $\lambda_i$s as the diagonal entries. Thus we have proven

***Proposition 2.3.*** Let **A** be a linear transformation on a vector space of finite dimension n. Suppose that **A** has n distinct eigenvalues $\lambda_1, ..., \lambda_n$. Then there is a basis **e** in which the matrix of components of **A** is diagonal, with diagonal entries $\lambda_1, ..., \lambda_n$.

Unfortunately, the story may be much more complicated when the eigenvalues fail to be distinct. Problem 2.16 shows that there are linear transformations **A** for which no basis **e** exists relative to which the matrix $\underline{A}^e$ of components is diagonal; such transformations clearly have eigenvalues with algebraic multiplicity greater than one. This state of affairs suggests an obvious question: if a linear transformation **A** refuses to have a diagonal matrix of components in *any* basis, what basis yields the simplest *non-diagonal* matrix of components for **A**? In a complex vector space, the answer to this question leads to the *Jordan canonical form*: there is a basis in which the matrix of components of **A**, while not necessarily diagonal, has zero entries in all positions *except* on the principal diagonal *and* on the diagonal immediately above the principal one. On the principal diagonal, the entries are the eigenvalues of **A**, and on the diagonal above, every entry is either zero or one. A precise statement of this theorem, together with a clear proof, may be found in Sections 18 and 19 of the book by Gel'fand [2.3] cited below. Because this is a theorem about *complex* vector spaces, we neither state nor prove it here, despite the fact that it is useful in a variety of contexts. For applications of the Jordan canonical form in the theory of ordinary differential equations, the reader is referred to the book by Hirsch and Smale [2.5] among the references listed below.

## References

[2.1] R. Bellman, *Introduction to Matrix Analysis*, McGraw-Hill, New York, 1960.

[2.2] R. Courant and D. Hilbert, *Methods of Mathematical Physics*, Volume 1, Interscience Press, New York, 1953.

[2.3] I.M. Gel'fand, *Lectures on Linear Algebra*, translated from the Russian by A. Shenitzer, Dover Publications, New York, 1989.

[2.4] P.R. Halmos, *Finite Dimensional Vector Spaces*, Second Edition, Van Nostrand-Reinhold, New York, 1958.

[2.5] M.W. Hirsch and S. Smale, *Differential Equations, Dynamical Systems and Linear Algebra*, Academic Press, Orlando, 1974.

## Problems

**2.1.** Let $\mathbf{p} = \alpha_0 + \alpha_1 t + \dots + \alpha_N t^N$ be a polynomial in the space $P_n$, $n = N + 1$. For every such **p**, let $\mathbf{q} = \mathbf{A}\mathbf{p}$ be the polynomial $\alpha_N + \alpha_0 t + \alpha_1 t^2 + \dots + \alpha_{N-1} t^N$. Show that the transformation **A** that carries **p** into **q** is linear and non-singular.

**2.2.** In the real vector space $\mathsf{R}_2$ of two-rowed columns of real numbers, consider the natural basis $\mathbf{e}_1 = \binom{1}{0}$, $\mathbf{e}_2 = \binom{0}{1}$, and let **A** and **B** be linear transformations whose matrices of components in this basis are given by

$$\underline{A}^{\mathbf{e}} = \begin{pmatrix} 1 & 2^{1/2} \\ 2^{1/2} & 1 \end{pmatrix}, \quad \underline{B}^{\mathbf{e}} = \begin{pmatrix} \cos\varphi & \sin\varphi \\ -\sin\varphi & \cos\varphi \end{pmatrix},$$

where $\varphi$ is a real number. (a) Find the eigenvalues $\lambda_1$, $\lambda_2$ of **A**. (b) Find the inverse of **B**. (c) If $\mathbf{C} = \mathbf{B}^{-1}\mathbf{A}\mathbf{B}$, show that $\varphi$ can be chosen so that $\underline{C}^{\mathbf{e}} = \begin{pmatrix} \lambda_1 & 0 \\ 0 & \lambda_2 \end{pmatrix}$.

**2.3.** Let $\mathsf{C}_0^\infty$ be the set of all real-valued functions f(t) that are defined and infinitely differentiable on the interval [0, 1] and vanish at t = 0. Let $\mathsf{C}_0^{\infty\prime}$ be a second collection of real functions f(t) whose definition is the same as that of $\mathsf{C}_0^\infty$ *except* for the additional restriction that all derivatives of each f in $\mathsf{C}_0^{\infty\prime}$ vanish at t = 0 as well. Suppose that $\mathsf{C}_0^\infty$ and $\mathsf{C}_0^{\infty\prime}$ are made into real vector spaces through the natural operations of addition of functions and multiplication of functions by a real number. (a) Verify that $\mathsf{C}_0^\infty$ and $\mathsf{C}_0^{\infty\prime}$ are indeed vector spaces. (b) Let **A**, **B** and **C** be defined on $\mathsf{C}_0^\infty$ as follows:

$$\mathbf{A}f(t) = \max_{0\le\tau\le t} f(\tau), \quad 0 \le t \le 1,$$
$$\mathbf{B}f(t) = f'(t) = \text{derivative of f at t}, \quad 0 \le t \le 1,$$
$$\mathbf{C}f(t) = \int_0^t f(\tau)\, d\tau, \quad 0 \le t \le 1.$$

Which of **A**, **B**, and **C** are linear transformations of $\mathsf{C}_0^\infty$ into itself? (c) How would the answer to this question change if **A**, **B**, and **C** were defined on $\mathsf{C}_0^{\infty\prime}$ instead of $\mathsf{C}_0^\infty$? (d) Does **C**, considered as a transformation on $\mathsf{C}_0^{\infty\prime}$, have an inverse? (e) Show that the function f(t) defined by

$$f(t) = \begin{cases} 0 & \text{for } t = 0, \\ e^{-1/t} & \text{for } 0 < t \le 1, \end{cases}$$

belongs to $\mathsf{C}_0^{\infty\prime}$.

**2.4.** Let **A** and **B** be a non-singular linear transformations on a vector space $\mathsf{R}$. Show that $\mathbf{C} = \mathbf{A}\mathbf{B}$ is also non-singular, and that its inverse is $\mathbf{C}^{-1} = \mathbf{B}^{-1}\mathbf{A}^{-1}$.

**2.5.** For each value of the real parameter $\alpha$, define a linear transformation

**A** on the vector space $\mathsf{R}_2$ by

$$\mathbf{y} = \begin{pmatrix} y_1 \\ y_2 \end{pmatrix} = \mathbf{Ax} = \begin{pmatrix} 1 & \alpha \\ 1 & 1 \end{pmatrix} \begin{pmatrix} x_1 \\ x_2 \end{pmatrix}.$$

For what value of $\alpha$ is **A** singular? When **A** is singular, find its null space and its range.

**2.6.** Consider the linear transformation **A** of (2.4) on the real space $\mathsf{R}_2$, and refer to Example 2.8. (a) Show that **A** has a one-dimensional invariant linear manifold if and only if the inequality (2.18) holds, and that when this inequality holds *strictly*, **A** has two linearly independent eigenvectors, and therefore two distinct one-dimensional invariant linear manifolds, corresponding to two distinct eigenvalues. (b) Show that when *equality* holds in (2.18), **A** has only one eigenvalue; its multiplicity as a zero of the characteristic polynomial is, of course, two. (c) Suppose it happens that $a_{12} = a_{21}$, and equality holds in (2.18). Show that in this case, **A** has two linearly independent eigenvectors, and therefore two distinct one-dimensional invariant linear manifolds, even though there is only one eigenvalue. (d) Suppose that $a_{11} = a_{22} = 1$, $a_{12} = 1$, $a_{21} = 0$. Show that in this case, there is only one eigenvalue *and* only one one-dimensional invariant linear manifold. Thus **A** *always* has two distinct one-dimensional invariant linear manifolds when it has two distinct eigenvalues; when there is only *one* eigenvalue, there may or may not be two distinct one-dimensional invariant linear manifolds. The number of linearly independent eigenvectors of **A** is therefore always two when **A** has two distinct eigenvalues, but this number may be either one or two when **A** has only one eigenvalue.

**2.7.** Let **A** be a linear transformation on an n-dimensional real vector space. Suppose that in a certain basis, the matrix $\underline{A}$ of components $a_{ij}$ is "upper triangular": $a_{ij} = 0$ for $i > j$. Find the eigenvalues of **A**. In the special case $n = 3$, find $\mathbf{A}^{-1}$ when **A** is non-singular.

**2.8.** Consider the integral equation (2.20), and suppose that the function $\mathbf{p}(t)$ satisfying (2.20) is not necessarily a polynomial, but *is* a real-valued function that is continuous for all $t \geq 0$. Are there values of $\lambda$ such that $\mathbf{p}(t)$ is not identically zero and (2.20) holds? Is the set of all such $\lambda$ the same as that determined in Example 2.9, where (2.20) was viewed as an equation in the space $\mathsf{P}_n$ of real polynomials of degree at most $n - 1$?

**2.9.** Let **R** be a linear transformation on an n-dimensional vector space, and suppose that **R** is such that $\mathbf{Re}_i = \mathbf{f}_i$, where $\mathbf{e}_1, ..., \mathbf{e}_n$ and $\mathbf{f}_1, ..., \mathbf{f}_n$ comprise two bases for the space. Show that **R** is non-singular.

**2.10.** Let **A** be a linear transformation on a vector space R of finite dimension n. If k = dim *R*(**A**) and m = dim *N*(**A**), show that k + m = n. Thus a *singular* transformation **A** collapses the vector space R onto onto a space of lesser dimension, namely *R*(**A**).

**2.11.** Let **A** and **B** be two linear transformations on a vector space R. Suppose they are related through a non-singular linear transformation **R** according to the formula $\mathbf{B} = \mathbf{R}^{-1}\mathbf{A}\mathbf{R}$. Show that if $\lambda$ is an eigenvalue of **A**, it is also an eigenvalue of **B**. What is the relation between the corresponding eigenvectors?

**2.12.** Let **B** be a linear transformation on a two-dimensional real vector space. Let **x** and **y** be linearly independent vectors, and suppose that

$$\mathbf{Bx} = \alpha\mathbf{x} + \beta\mathbf{y}, \quad \mathbf{By} = \gamma\mathbf{x} + \delta\mathbf{y},$$

so that the matrix of components of **B** in the basis **x**, **y** is $\underline{\mathrm{B}}^{(1)} = \begin{pmatrix} \alpha & \gamma \\ \beta & \delta \end{pmatrix}$. Let **R** be a non-singular tensor, and set $\mathbf{u} = \mathbf{Rx}$, $\mathbf{v} = \mathbf{Ry}$. Let $\underline{\mathrm{B}}^{(2)}$ be the matrix of components of **B** in the new basis **u**, **v**. By the change-of-basis formula in the form (2.36),

$$\underline{\mathrm{R}}\underline{\mathrm{B}}^{(2)} = \underline{\mathrm{B}}^{(1)}\underline{\mathrm{R}}, \quad (*)$$

where $\underline{\mathrm{R}}$ is the matrix of components of **R** in the basis **x**, **y**. Suppose that the new basis **u**, **v** can be chosen such that

$$\underline{\mathrm{B}}^{(2)} = \begin{pmatrix} \mu & \nu \\ -\nu & \mu \end{pmatrix}, \quad (**)$$

where $\mu$, $\nu$ are real, with $\nu > 0$. (a) Using properties of the trace and the determinant, show that $\mu$ and $\nu$ must necessarily satisfy

$$\mu = (\alpha + \delta)/2, \quad \mu^2 + \nu^2 = \alpha\delta - \beta\gamma. \quad (***)$$

(b) Show that (***), along with $\nu > 0$, implies that $(\delta - \alpha)^2/4 + \beta\gamma < 0$. (c) Show that if **B** has no one-dimensional invariant subspace, then the inequality of part (b) is indeed satisfied by $\alpha$, $\beta$, $\gamma$ and $\delta$. (d) Let $\underline{\mathrm{R}} = \begin{pmatrix} \mathrm{p} & \mathrm{q} \\ \mathrm{r} & \mathrm{s} \end{pmatrix}$; assuming that $\underline{\mathrm{B}}^{(2)}$ has the form (**), reduce (*) to a homogeneous linear system of four equations for p, q, r, and s. (e) Show that if (***) holds, then this system has solutions such that det $\mathbf{R} \neq 0$, so that there is indeed a basis **u**, **v** for which (**) holds. (f) Show finally that **u** and **v** satisfy

$$\mathbf{Bu} = \mu\mathbf{u} - \nu\mathbf{v}, \quad \mathbf{Bv} = \nu\mathbf{u} + \mu\mathbf{v}. \quad (****)$$

**2.13.** Let M be a two-dimensional invariant linear manifold for a linear transformation **A**, and suppose that M contains no *one*-dimensional invariant linear manifold for **A**. Identify the restriction of **A** to M with the transformation **B** of the preceding problem, and use the results of that problem to prove that there are linearly independent vectors **u** and **v** in M such that

$$\mathbf{Au} = \mu\mathbf{u} - \nu\mathbf{v}, \quad \mathbf{Av} = \nu\mathbf{u} + \mu\mathbf{v}.$$

Thus in seeking two dimensional invariant linear manifolds, there is no loss of generality in assuming that their generators **x**, **y** satisfy the special version (2.44) of (2.24). For a transformation **A** on a finite-dimensional real space, it follows that *every* two-dimensional invariant linear manifold for **A** corresponds to a complex zero of the characteristic polynomial $P_A(\lambda)$.

**2.14.** Let **A** be a linear transformation on a real vector space R, and suppose it has a two-dimensional invariant linear manifold M generated by a linearly independent set of two vectors **x** and **y** satisfying (2.44) with $\nu \neq 0$. Show that there is no real number $\lambda$ such that $\mathbf{Az} = \lambda\mathbf{z}$ for some non-null vector **z** in M. Thus there is no *one*-dimensional invariant linear manifold hiding within M; in this sense, M is *irreducible.*

**2.15.** Show that Tr (**AB**) = Tr (**BA**) for any two linear transformations **A** and **B**.

**2.16.** Let **A** be a linear transformation on a two-dimensional real vector space. Suppose there is a basis in which the matrix of components of **A** is $\begin{pmatrix} 1 & 1 \\ 0 & 1 \end{pmatrix}$. Show that there is *no* basis in which the matrix of components of **A** is diagonal.

**2.17.** Let **A** be a linear transformation on an n-dimensional vector space. Let $\mathbf{e}_1, ..., \mathbf{e}_k$ be eigenvectors of **A**, and assume that the corresponding eigenvalues $\lambda_1, ..., \lambda_k$ are distinct. Prove that the set $\{\mathbf{e}_1, ..., \mathbf{e}_k\}$ is linearly independent by the following inductive argument. (a) Observe that the result is true if $k = 1$. (b) Assume the result is true for $k - 1$ vectors. Suppose that it fails to be true for k vectors. Then there are k scalars $\alpha_1, ..., \alpha_k$, not all zero, such that $\alpha_1\mathbf{e}_1 + \alpha_2\mathbf{e}_2 + ... + \alpha_k\mathbf{e}_k = \mathbf{o}$ (*). Operating on both sides of this equation with **A** yields $\alpha_1\lambda_1\mathbf{e}_1 + \alpha_2\lambda_2 \mathbf{e}_2 + ... + \alpha_k\lambda_k\mathbf{e}_k = \mathbf{o}$ (**). (c) Show that (*) and (**) imply that $\alpha_1(\lambda_1 - \lambda_k)\mathbf{e}_1 + \alpha_2(\lambda_2 - \lambda_k)\mathbf{e}_2 + ... + \alpha_{k-1}(\lambda_{k-1} - \lambda_k)\mathbf{e}_{k-1} = \mathbf{o}$. (d) Thus show that $\alpha_1(\lambda_1 - \lambda_k) = \alpha_2(\lambda_2 - \lambda_k) = ... = \alpha_{k-1}(\lambda_{k-1} - \lambda_k) = 0$. (e) Use the distinctness of the eigenvalues to infer that $\alpha_1 = \alpha_2 = ... = \alpha_{k-1} = 0$. (e) Use (*) to show that $\alpha_k = 0$, contradicting the fact that not all k $\alpha_i$s vanish. Thus $\mathbf{e}_1, ..., \mathbf{e}_k$ are linearly independent.

**2.18.** Let **A** be a linear transformation on a complex vector space R of finite dimension n. Assume that **A** has n distinct eigenvalues $\lambda_1, ..., \lambda_n$. Then by Proposition 2.3, there is a basis **e** in which the matrix $\underline{A}^e$ is diagonal, with the $\lambda_i$s as diagonal entries. (a) Let k be a positive integer. Show that the linear transformation $\mathbf{A}^k$, by which is meant the k-fold iterate of **A**, also has a diagonal matrix in the basis **e**, diagonal entries $\lambda_i{}^k$, i = 1, ..., n. (b) Let p(**A**) be any polynomial in **A** with coefficients in the scalar field for R. Then p(**A**) also has a diagonal matrix of components in the basis **e**. (c) Use these results to establish the *Cayley-Hamilton theorem* in the case of a linear transformation with distinct eigenvalues; *If* **A** *is a linear transformation on a finite dimensional complex vector space, and if* $P_A(\lambda)$ *is the characteristic polynomial of* **A**, *then* **A** *satisfies* $P_A(\mathbf{A}) = \mathbf{O}$. This theorem remains true if the eigenvalues are not distinct; see Section 1, Chapter 6 of Hirsch and Smale [2.5].

**2.19.** Let R be an n-dimensional complex vector space. Use the Cayley-Hamilton theorem (see the preceding problem) to show that the inverse of a non-singular linear transformation **A** on R can be expressed as a polynomial of degree n − 1 in **A**.

**2.20** Let **A** and **B** be linear transformations of a *two-dimensional* complex vector space into itself. Set **C** = **AB** − **BA**. Use the Cayley-Hamilton theorem of Problem 2.18 to show that $\mathbf{C}^2$ is a scalar multiple of the identity.

CHAPTER 3

# FINITE-DIMENSIONAL EUCLIDEAN SPACES AND CARTESIAN TENSORS

We now restrict our attention to finite-dimensional real Euclidean spaces, which are by far the most important finite-dimensional spaces for applications in mechanics. Linear transformations on such a space are called *Cartesian tensors*, or just *tensors* for brevity. The effort in this chapter is directed toward understanding the geometric properties of tensors; these properties are responsible for the utility of the theory in the applications to be discussed later.

## Preliminaries

We now consider a real vector space R of finite dimension n that has been rendered a real Euclidean space by means of a scalar product meeting all the requirements set forth in Chapter 1. If **A** is a linear transformation on such a space, i.e., a tensor on R, and if $\mathbf{e}_1, \mathbf{e}_2, \ldots, \mathbf{e}_n$ is an orthonormal basis for R, taking the scalar product of both sides of (2.25) with $\mathbf{e}_i$ shows that the components $a_{ij}$ of **A** in **e** are given by

$$a_{ij} = (\mathbf{e}_i, \mathbf{A}\mathbf{e}_j) \tag{3.1}$$

This provides a convenient way to determine the components of a tensor in an orthonormal basis.

The first new notion of importance in this setting is that of the *transpose* of a linear transformation—or tensor—**A** on R. **A** is said to have a *transpose* **B** if there is a tensor **B** with the property that

$$(\mathbf{A}\mathbf{x}, \mathbf{y}) = (\mathbf{x}, \mathbf{B}\mathbf{y}) \quad \text{for all } \mathbf{x}, \mathbf{y} \text{ in R.} \tag{3.2}$$

It is easy to see that (3.2) holds if and only if the equality is satisfied for all *unit* vectors **x**, **y** in R, that is for all vectors such that $|\mathbf{x}| = |\mathbf{y}| = 1$. For this latter class of vectors, (3.2) says that the tensor **B** has the property that the projection of its output **By** on **x** is the same as the projection of the output **Ax** on **y**. It is not clear *a priori* why this notion has the far-reaching consequences that it will prove to have; indeed, at this point it is not even clear that, given an **A**, there *is* a **B** satisfying (3.2).

We first show that, for each **A**, there is *at most* one **B** satisfying (3.2). Suppose there were two such **B**s, say $\mathbf{B}_1$ and $\mathbf{B}_2$. Then from (3.2) it would immediately follow that

$$(\mathbf{x}, (\mathbf{B}_1 - \mathbf{B}_2)\mathbf{y}) = 0 \text{ for all } \mathbf{x}, \mathbf{y} \text{ in } \mathsf{R}. \tag{3.3}$$

Let **y** be fixed; according to (3.3), the vector $(\mathbf{B}_1 - \mathbf{B}_2)\,\mathbf{y}$ is orthogonal to *every* vector **x**, so that—according to the story set out in Chapter 1—it must vanish. It follows that $(\mathbf{B}_1 - \mathbf{B}_2)\,\mathbf{y} = \mathbf{o}$ *for every* **y**, and hence that $\mathbf{B}_1 - \mathbf{B}_2 = \mathbf{O}$, or $\mathbf{B}_1 = \mathbf{B}_2$.

But is there such a tensor **B**? We now answer this question with the help of the notion of a *linear function* on R. Such a function—say $\psi$—associates with each vector **x** in R a real number $\psi(\mathbf{x})$ in such a way that $\psi(\alpha\mathbf{x} + \beta\mathbf{y}) = \alpha\,\psi(\mathbf{x}) + \beta\psi(\mathbf{y})$ for every **x**, **y** in R and for every pair of real numbers $\alpha$, $\beta$. Problem 3.1 asserts that, if R is finite dimensional, then for any such linear function $\psi$, there is a vector **h** in R such that $\psi(\mathbf{x}) = (\mathbf{h}, \mathbf{x})$ for all **x** in R. This says that the only linear function of **x** in R is the scalar product of **x** with some fixed vector **h** in R; different $\psi$s have different **h**s. We apply this in the situation at hand by setting $\psi_{\mathbf{y}}(\mathbf{x}) = (\mathbf{Ax}, \mathbf{y})$, thinking of **y** as momentarily fixed with **x** free to run around in R. Clearly $\psi_{\mathbf{y}}$ is a real-valued linear function on R, and so by the result cited above, there is an **h** in R—presumably dependent upon the choice of the fixed vector **y**, so that $\mathbf{h} = \mathbf{h}(\mathbf{y})$—such that $\psi_{\mathbf{y}}(\mathbf{x}) = (\mathbf{h}(\mathbf{y}), \mathbf{x})$. Thus $(\mathbf{Ax}, \mathbf{y}) = (\mathbf{h}(\mathbf{y}), \mathbf{x})$. Because **A** is linear, the properties of the scalar product allow us to observe immediately that $(\mathbf{h}(\alpha\mathbf{y} + \beta\mathbf{z}), \mathbf{x}) = (\alpha\mathbf{h}(\mathbf{y}) + \beta\mathbf{h}(\mathbf{z}), \mathbf{x})$ for every triplet **x**, **y**, and **z** of vectors in R. Thus $\mathbf{h}(\alpha\mathbf{y} + \beta\mathbf{z}) - \alpha\mathbf{h}(\mathbf{y}) - \beta\mathbf{h}(\mathbf{z})$ is orthogonal to every vector **x** in R and must therefore vanish. It follows that the transformation on R that carries **y** to **h**(**y**) is a linear one: thus there is a tensor **B** such that $\mathbf{h}(\mathbf{y}) = \mathbf{By}$. In other words, $(\mathbf{Ax}, \mathbf{y}) = (\mathbf{By}, \mathbf{x})$ for every pair of vectors **x** and **y** in R. This was what we set out to show.

Given an **A**, we call the unique tensor **B** associated with **A** through the above argument the *transpose* of **A**, and we write $\mathbf{B} = \mathbf{A}^{\mathrm{T}}$. It is not hard to see that $(\mathbf{A}^{\mathrm{T}})^{\mathrm{T}} = \mathbf{A}$, and that if **A** and **C** are any two tensors, then $(\mathbf{AC})^{\mathrm{T}} = \mathbf{C}^{\mathrm{T}}\mathbf{A}^{\mathrm{T}}$; see Problem 3.2.

***Example 3.1.*** *The transpose in* $\mathsf{R}_n$. Let $\mathsf{R}_n$ be the real Euclidean space of columns **x** of n real numbers $x_1, ..., x_n$, with the scalar product of two vectors given by

$$(\mathbf{x}, \mathbf{y}) = x_i\, y_i. \tag{3.4}$$

Let **A** be the linear transformation on $\mathsf{R}_n$ defined by

$$\mathbf{Ax} = \begin{pmatrix} a_{11} & a_{12} & \dots & a_{1n} \\ a_{21} & a_{22} & \dots & a_{2n} \\ \dots & \dots & \dots & \dots \\ a_{n1} & a_{n2} & \dots & a_{nn} \end{pmatrix} \begin{pmatrix} x_1 \\ x_2 \\ \dots \\ x_n \end{pmatrix} = \begin{pmatrix} a_{1j} & x_j \\ a_{2j} & x_j \\ \dots & \dots \\ a_{nj} & x_j \end{pmatrix}, \tag{3.5}$$

where the $a_{ij}$ are real numbers. Let **B** be defined by an equation exactly like (3.5), but with $a_{ij}$ replaced by real numbers $b_{ij}$. Then by direct calculation,

$$(\mathbf{Ax}, \mathbf{y}) - (\mathbf{x}, \mathbf{By}) = a_{ij}\, x_j y_i - b_{ij} y_j x_i = (a_{ij} - b_{ji})\, x_j y_i; \tag{3.6}$$

it follows that $\mathbf{B} = \mathbf{A}^T$ if and only if $b_{ij} = a_{ji}$. Thus in $\mathsf{R}_n$, the transpose of **A** is obtained by interchanging the rows and columns of the $n \times n$ matrix in (3.5).

***Example 3.2.*** *The tensor product of two vectors.* Let $\mathsf{R}$ be any n-dimensional real Euclidean space, and let **a** and **b** be fixed vectors in $\mathsf{R}$. We define a tensor **C** by specifying its action on an arbitrary vector **x** as follows:

$$\mathbf{C}\,\mathbf{x} = (\mathbf{b}, \mathbf{x})\, \mathbf{a} \text{ for every } \mathbf{x} \text{ in } \mathsf{R}. \tag{3.7}$$

Thus **C** is a tensor whose output when fed **x** is a vector in the direction of **a** (or possibly $-\mathbf{a}$), but whose length depends on both **x** and **b** through the scalar product $(\mathbf{b}, \mathbf{x})$. The tensor **C** is called the *tensor product* of **a** and **b**, in that order, and we write $\mathbf{C} = \mathbf{a}\otimes\mathbf{b}$. The tensor product is useful in a variety of applications, some of which we shall encounter later. Observe that $\mathbf{C} = \mathbf{a}\otimes\mathbf{b}$ and $\mathbf{D} = \mathbf{b}\otimes\mathbf{a}$ are in general different tensors; indeed, from the definition (3.7), it follows that

$$(\mathbf{Cx}, \mathbf{y}) - (\mathbf{x}, \mathbf{Dy}) = (\mathbf{b}, \mathbf{x})(\mathbf{a}, \mathbf{y}) - (\mathbf{a}, \mathbf{y})(\mathbf{b}, \mathbf{x}) = 0, \tag{3.8}$$

so that $\mathbf{D} = \mathbf{C}^T$, or

$$(\mathbf{a}\otimes\mathbf{b})^T = \mathbf{b}\otimes\mathbf{a}. \tag{3.9}$$

In particular, any scalar multiple of $\mathbf{a}\otimes\mathbf{a}$ is a symmetric tensor. Incidentally, it is useful to record here the components of $\mathbf{a}\otimes\mathbf{b}$ in an orthonormal basis $\mathbf{e}_1, ..., \mathbf{e}_n$; from (3.1) and (3.7), the components $c_{ij}$ in **e** of **C** are $c_{ij} = (\mathbf{e}_i, \mathbf{Ce}_j) = (\mathbf{e}_i, \mathbf{a})\, (\mathbf{e}_j, \mathbf{b})$. Thus if $a_k = (\mathbf{a}, \mathbf{e}_k)$ and $b_k = (\mathbf{b}, \mathbf{e}_k)$ are the components of the vectors **a** and **b** in the basis **e**, we have shown that $c_{ij} = a_i\, b_j$, or—in suggestive notation—$(\mathbf{a}\otimes\mathbf{b})_{ij} = a_i\, b_j$.

Let **A** be a tensor on an n-dimensional real Euclidean space. Let $\mathbf{e}_1, \mathbf{e}_2, ..., \mathbf{e}_n$ be any orthonormal basis, and let $a_{ij} = (\mathbf{e}_i, \mathbf{Ae}_j)$ be the components of $\mathbf{A}$ in $\mathbf{e}$. Let $\mathbf{B} = \mathbf{A}^T$; the components $b_{ij}$ in $\mathbf{e}$ of $\mathbf{B}$ are given by $b_{ij} = (\mathbf{e}_i, \mathbf{B}\mathbf{e}_j) = (\mathbf{e}_i, \mathbf{A}^T\mathbf{e}_j) = (\mathbf{A}\mathbf{e}_i, \mathbf{e}_j) = a_{ji}$. Thus the matrix $\underline{B}$ of components of $\mathbf{B}$ in $\mathbf{e}$ is the matrix-transpose of the matrix $\underline{A}$ of components of $\mathbf{A}$ in $\mathbf{e}$. This result is in general *not* true in a basis that is not orthonormal (see Problem 3.8) and is therefore *not* a basis-independent statement.

## Symmetric Tensors

A tensor $\mathbf{A}$ is *symmetric* if $\mathbf{A}^T = \mathbf{A}$, *skew-symmetric* if $\mathbf{A}^T = -\mathbf{A}$. For symmetric tensors, one clearly has $(\mathbf{Ax}, \mathbf{y}) = (\mathbf{x}, \mathbf{Ay})$ for all vectors $\mathbf{x}$, $\mathbf{y}$, while for skew-symmetric ones, $(\mathbf{Ax}, \mathbf{y}) = -(\mathbf{x}, \mathbf{Ay})$. In view of the discussion in the preceding paragraph, the components $a_{ij}$ of $\mathbf{A}$ *in an orthonormal basis* satisfy $a_{ij} = a_{ji}$ if $\mathbf{A}$ is symmetric, $a_{ij} = -a_{ji}$ if $\mathbf{A}$ is skew-symmetric. In particular, each diagonal component $a_{ii}$ (no sum on i) of a skew-symmetric tensor must necessarily vanish.

The symmetric tensors are among the most felicitous of mathematical creatures: they have fine properties, are easy to work with, and are of great importance in applications. We explore next the nature of this cooperative class of linear transformations.

***Proposition 3.1.*** Let $\mathbf{A}$ be a symmetric tensor. Then every zero of the characteristic polynomial $P_{\mathbf{A}}(\lambda)$ is real.

To prove this proposition, we begin by assuming that $\lambda = \mu + i\nu$, $\nu \neq 0$, is a complex zero of $P_{\mathbf{A}}(\lambda)$. According to the discussion at the end of the preceding chapter, there are then two linearly independent vectors $\mathbf{x}$ and $\mathbf{y}$ such that

$$\mathbf{Ax} = \mu\mathbf{x} - \nu\mathbf{y}, \quad \mathbf{Ay} = \nu\mathbf{x} + \mu\mathbf{y}. \tag{3.10}$$

From these equations, it follows that $(\mathbf{y}, \mathbf{Ax}) - (\mathbf{x}, \mathbf{Ay}) = -\nu\,(|\mathbf{x}|^2 + |\mathbf{y}|^2)$. But because $\mathbf{A}$ is symmetric, $(\mathbf{y}, \mathbf{Ax}) - (\mathbf{x}, \mathbf{Ay}) = 0$. Since $\mathbf{x}$ and $\mathbf{y}$ are linearly independent and thus not null, and because $\nu \neq 0$, we have a contradiction. Thus we conclude that $\nu = 0$, so that every zero $\lambda$ of $P_{\mathbf{A}}(\lambda)$ is necessarily real.

***Proposition 3.2.*** Let $\mathbf{A}$ be a symmetric tensor. Then there are n orthonormal eigenvectors $\mathbf{e}_1, ..., \mathbf{e}_n$ of $\mathbf{A}$ corresponding to n eigenvalues $\lambda_1, ..., \lambda_n$: $\mathbf{A}\mathbf{e}_i = \lambda_i\mathbf{e}_i$ (no sum on i).

To establish this result, we begin by noting that the preceding proposition assures the existence of at least one one-dimensional invariant linear manifold for **A**, say $\mathsf{M}_1$, corresponding to the real zero of $P_{\mathbf{A}}(\lambda)$, say $\lambda_1$, promised by that proposition. Define the set $\mathsf{M}_1^{\perp}$ of all vectors "perpendicular to $\mathsf{M}_1$" by

$$\mathsf{M}_1^{\perp} = \{\mathbf{x} \text{ in } \mathsf{R} \mid (\mathbf{x}, \mathbf{y}) = 0 \text{ for every } \mathbf{y} \text{ in } \mathsf{M}_1\}; \tag{3.11}$$

the linear manifold $\mathsf{M}_1^{\perp}$ is called the *orthogonal complement* of $\mathsf{M}_1$. Let **x** be in $\mathsf{M}_1^{\perp}$, and consider $(\mathbf{Ax}, \mathbf{y})$, where **y** is any vector in $\mathsf{M}_1$. Clearly $(\mathbf{Ax}, \mathbf{y}) = (\mathbf{x}, \mathbf{Ay})$ by the symmetry of **A**; moreover, since **y** is in $\mathsf{M}_1$, $\mathbf{Ay} = \lambda_1\mathbf{y}$, so $(\mathbf{Ax}, \mathbf{y}) = \lambda_1(\mathbf{x}, \mathbf{y}) = 0$. Thus **Ax** is in $\mathsf{M}_1^{\perp}$ whenever **x** is in $\mathsf{M}_1^{\perp}$, which is therefore also an invariant linear manifold for **A**. Consider now the restriction of **A** to the new vector space $\mathsf{M}_1^{\perp}$, whose dimension is $n - 1$. Appealing again to the preceding proposition, we infer that, within $\mathsf{M}_1^{\perp}$, there is a one-dimensional invariant linear manifold for **A**, call it $\mathsf{M}_2$, corresponding to a zero, say $\lambda_2$, of $P_{\mathbf{A}}(\lambda)$. Again, the set $\mathsf{M}_2^{\perp}$ of all vectors *lying in the space* $\mathsf{M}_1$ and perpendicular to $\mathsf{M}_2$ is an invariant linear manifold for **A** of dimension $n - 2$. We repeat this process until we arrive at the last one-dimensional invariant linear manifold $\mathsf{M}_n$ for **A**, corresponding to the $n^{th}$ zero $\lambda_n$ of $P_{\mathbf{A}}(\lambda)$. Let $\mathbf{e}_1, \mathbf{e}_2, ..., \mathbf{e}_n$ be vectors of unit length generating $\mathsf{M}_1, \mathsf{M}_2, ..., \mathsf{M}_n$, respectively. Then by construction, the $\mathbf{e}_i$s are clearly orthogonal, and $\mathbf{Ae}_i = \lambda_i\mathbf{e}_i$, no sum on i. This was what we wished to show.

It should be noted that, in the argument given above, nothing was assumed about the distinctness of the eigenvalues $\lambda_i$ of **A**, some or all of which might be repeated zeros of $P_{\mathbf{A}}(\lambda)$.

It is useful to describe the conclusion of the last proposition in two alternate ways. First, we note that the formula $\mathbf{Ae}_i = \lambda_i\mathbf{e}_i$ (no sum) immediately implies that the matrix of components $\underline{A}^e$ of **A** in the special orthonormal basis $\mathbf{e}_1, \mathbf{e}_2, ..., \mathbf{e}_n$ has the following diagonal form:

$$\underline{A}^e = \begin{pmatrix} \lambda_1 & 0 & \cdots & \cdots & 0 \\ 0 & \lambda_2 & 0 & \cdots & 0 \\ \cdots & \cdots & \cdots & \cdots & \cdots \\ 0 & \cdots & \cdots & 0 & \lambda_n \end{pmatrix} \tag{3.12}$$

An orthonormal basis with this property is said to be a *principal basis* for the symmetric tensor **A**. A symmetric tensor may have many principal bases; for the identity tensor **1**, for example, *any* set of n orthonormal vectors comprises a principal basis. Such bases are extremely useful both for calculations and for understanding the physical or geometrical meaning of the action of a symmetric tensor. We shall return to this point shortly.

Still another way to describe the result of the last proposition involves the so-called *spectral formula* for a symmetric tensor **A**. If $\lambda_1, \lambda_2, ..., \lambda_n$ are the eigenvalues of **A**, with $\mathbf{e}_1, \mathbf{e}_2, ..., \mathbf{e}_n$ a set of corresponding orthonormal eigenvectors, one has the following handsome result:

$$\mathbf{A} = \lambda_1\, \mathbf{e}_1 \otimes \mathbf{e}_1 + \lambda_2\, \mathbf{e}_2 \otimes \mathbf{e}_2 + \ldots + \lambda_n\, \mathbf{e}_n \otimes \mathbf{e}_n. \tag{3.13}$$

To prove (3.13), we let $\mathbf{x}$ be any vector in R, and we represent $\mathbf{x}$ in terms of the principal basis $\mathbf{e}$ for $\mathbf{A}$: $\mathbf{x} = \xi_i\, \mathbf{e}_i$, where $\xi_i = (\mathbf{x}, \mathbf{e}_i)$. On the one hand, because the $\mathbf{e}_i$s are eigenvectors, $\mathbf{Ax} = \xi_i \mathbf{A}\mathbf{e}_i = \sum_{i=1}^{n} \lambda_i \xi_i\, \mathbf{e}_i$. On the other hand, $\sum_{i=1}^{n} \lambda_i\, (\mathbf{e}_i \otimes \mathbf{e}_i)\, \mathbf{x} = \sum_{i=1}^{n} \lambda_i\, (\mathbf{x}, \mathbf{e}_i)\, \mathbf{e}_i = \sum_{i=1}^{n} \lambda_i \xi_i\, \mathbf{e}_i$. Thus when given an arbitrary input $\mathbf{x}$ in R, the tensors on both sides of (3.13) produce the same output, and hence coincide, establishing the *spectral formula* (3.13).

In the preceding chapter, it was observed that the geometric and algebraic multiplicities of an eigenvalue of a linear transformation on a finite dimensional vector space need not coincide. As shown in the following proposition, the two multiplicities are *always* the same for a symmetric tensor.

---

***Proposition 3.3.*** Let $\mathbf{A}$ be a symmetric tensor, and let $\lambda_0$ be an eigenvalue of $\mathbf{A}$. Then the algebraic and geometric multiplicities of $\lambda_0$ are equal.

---

To prove this, let k and m be the geometric and algebraic multiplicities of $\lambda_0$, respectively, and let M be the k-dimensional invariant linear manifold for $\mathbf{A}$ spanned by the eigenvectors corresponding to $\lambda_0$. Then the restriction $\mathbf{A}_0$ of $\mathbf{A}$ to M is a symmetric linear transformation of M into itself; moreover, $\mathbf{A}_0\mathbf{x} = \lambda_o\mathbf{x}$ for *every* $\mathbf{x}$ in M, so that $\mathbf{A}_0 = \lambda_0 \mathbf{1}_k$, where $\mathbf{1}_k$ is the k-dimensional identity tensor. Thus the characteristic polynomial of $\mathbf{A}_0$ is given by $P_{\mathbf{A}_0}(\lambda) = (\lambda_0 - \lambda)^k$. Next, let $\mathbf{A}_1$ be the restriction of $\mathbf{A}$ to the orthogonal complement $M^\perp$ of M; $\mathbf{A}_1$ is a symmetric linear transformation of the (n − k)-dimensional space $M^\perp$ into itself. Thus there are n − k orthonormal eigenvectors $\mathbf{e}_{k+1}, \ldots, \mathbf{e}_n$ of $\mathbf{A}_1$ (and hence of $\mathbf{A}$ as well) in $M^\perp$ corresponding to the respective eigenvalues $\lambda_{k+1}, \ldots, \lambda_n$. Thus $P_{\mathbf{A}_1}(\lambda) = (\lambda_{k+1} - \lambda) \cdots (\lambda_n - \lambda)$; it is clear that $\lambda_i \neq \lambda_0$ for $i = k + 1, \ldots, n$. It is easy to show that $P_{\mathbf{A}}(\lambda) = P_{\mathbf{A}_0}(\lambda)\, P_{\mathbf{A}_1}(\lambda)$; convince yourself of this. It follows that the algebraic multiplicity m of $\lambda_0$ is the algebraic multiplicity of $\lambda_0$ as an eigenvalue of $\mathbf{A}_0$, which is k. Thus m = k, and the proposition is established.

One often encounters the question of whether two symmetric tensors $\mathbf{A}$ and $\mathbf{B}$ have a *common* principal basis. If the answer is affirmative, then the matrices of components of $\mathbf{A}$ and $\mathbf{B}$ in such a basis are *both* diagonal. The following proposition describes precisely the conditions under which this occurs.

---

***Proposition 3.4.*** Let $\mathbf{A}$ and $\mathbf{B}$ be symmetric tensors. Then $\mathbf{A}$ and $\mathbf{B}$ have a common principal basis if and only if $\mathbf{A}$ and $\mathbf{B}$ commute: $\mathbf{AB} = \mathbf{BA}$.

---

To demonstrate this result, we begin by observing that, if **A** and **B** *do* have a common principal basis, then their respective matrices A and B in this basis are both diagonal, so certainly AB = BA and therefore **AB** = **BA**. Thus commutativity of **A** and **B** is necessary for the existence of a common principal basis. To show that it is also sufficient, assume that **AB** = **BA**, and let **b** be an eigenvector of **B** corresponding to the eigenvalue $\beta$ of multiplicity k. Thus **B** has a k-dimensional invariant linear manifold M containing an orthonormal basis $\mathbf{b}_1, ..., \mathbf{b}_k$ such that $\mathbf{B}\mathbf{b}_j = \beta\mathbf{b}_j, j = 1, ..., k$. Let **f** be any vector in M. Then $\mathbf{f} = \sum_{j=0}^{k}\gamma_j\mathbf{b}_j$ for suitable scalars $\gamma_j$. Then $\mathbf{Bf} = \beta\mathbf{f}$, whence $\mathbf{ABf} = \beta\,\mathbf{Af}$. Since A and **B** commute, this means that $\mathbf{B}(\mathbf{Af}) = \beta\,(\mathbf{Af})$, so that **Af** is either null or an eigenvector of **B**; in either event, **Af** is in M. Hence M is an invariant linear manifold for **A**, and the restriction of **A** to M is a symmetric linear transformation of M into itself. It follows that **A** has k orthonormal eigenvectors $\mathbf{a}_1, ..., \mathbf{a}_k$ in M corresponding to eigenvalues $\alpha_1, ..., \alpha_k$: $\mathbf{A}\mathbf{a}_j = \alpha_j\mathbf{a}_j$ (no sum on j). Since the $\mathbf{a}_j$s are in M, they are also eigenvectors of **B**, all associated with the eigenvalue $\beta$: $\mathbf{B}\mathbf{a}_j = \beta\mathbf{a}_j$. In the basis **a**, the matrices of components of the restrictions of **A** and **B** to the k-dimensional space M are therefore diagonal, with respective entries $\alpha_1, ..., \alpha_k$ and $\beta, \beta, ..., \beta$ along the principal diagonal.

To complete the proof of the proposition, one applies the argument carried out above to each of the *distinct* eigenvalues $\beta_i$ of **B**, arriving at an orthonormal basis $\mathbf{a}_1^{(i)}, ..., \mathbf{a}_{k_i}^{(i)}$ for the corresponding invariant linear manifold $M_i$; here $k_i$ is the multiplicity of $\beta_i$ and therefore the dimension of $M_i$. The subscript i runs through values 1, ..., p, where p is the number of distinct eigenvalues of **B**. In the basis $\mathbf{a}^{(i)}$, the $k_i \times k_i$ matrices $A^{(i)}$ and $B^{(i)}$ of components of the restrictions of **A** and **B** to $M_i$ are diagonal. Taken altogether, the bases $\mathbf{a}^{(1)}, ..., \mathbf{a}^{(p)}$ form an orthonormal basis for the entire space R in which the matrices of components A and B are made up of the respective blocks $A^{(i)}$ and $B^{(i)}$, and therefore are themselves diagonal. The basis arrived at in this way is clearly a common principal basis for both **A** and **B**, and the proposition is established.

If all n of the eigenvalues $\lambda_i$ of a symmetric tensor **A** are *positive*, **A** is said to be *positive definite*; if all $\lambda_i$ are non-negative, **A** is *non-negative definite*. Negative definite and non-positive definite symmetric tensors are defined analogously. Note that these terms apply only to *symmetric* tensors. The scalar-valued function of **x** on R defined by $f(\mathbf{x}) = (\mathbf{Ax}, \mathbf{x})$ is called the *quadratic form* associated with the symmetric tensor **A**. Such forms often arise in applications; Problem 3.9 shows that $f(\mathbf{x}) > 0$ for every $\mathbf{x} \neq \mathbf{o}$ if and only if **A** is positive definite in the sense defined above. Thus the positive definiteness of the symmetric tensor **A** is equivalent to the positivity of the quadratic form $f(\mathbf{x}) = (\mathbf{Ax}, \mathbf{x})$ for all non-null **x**.

We note in passing that (3.12) shows that, if **A** is symmetric, then det $\mathbf{A} = \lambda_1\lambda_2 \ldots \lambda_n$; in particular, a positive definite, symmetric tensor has positive determinant and is thus non-singular.

If **A** is a positive definite, symmetric tensor, it has a uniquely defined positive definite "square root"; that is, there is a unique positive definite, symmetric tensor **B** such that $\mathbf{B}^2 = \mathbf{A}$. To show this, we first note that if **e** is an eigenvector of **A** corresponding to the eigenvalue $\lambda$, and if there *is* such a tensor **B**, then $\mathbf{B}^2\,\mathbf{e} = \lambda\,\mathbf{e}$. Since $\lambda$ is necessarily positive, we may rewrite this last assertion as $(\mathbf{B} + \lambda^{1/2}\,\mathbf{1})(\mathbf{B} - \lambda^{1/2}\,\mathbf{1})\,\mathbf{e} = \mathbf{o}$. Set $\mathbf{f} = (\mathbf{B} - \lambda^{1/2}\mathbf{1})\,\mathbf{e}$. Then $\mathbf{B}\,\mathbf{f} + \lambda^{1/2}\,\mathbf{f} = \mathbf{o}$. But if **f** were different from **o**, **B** would have the negative eigenvalue $-\lambda^{1/2}$, which cannot happen because **B** is assumed to be positive definite. Thus necessarily $\mathbf{f} = \mathbf{o}$, or $\mathbf{B}\,\mathbf{e} = \lambda^{1/2}\,\mathbf{e}$. It follows that if $\lambda_1, \ldots, \lambda_n$ and $\mathbf{e}_1, \ldots, \mathbf{e}_n$ are the eigenvalues and corresponding eigenvectors of **A**, then **B**, if it exists, must have eigenvalues $\lambda_1{}^{1/2}, \lambda_2{}^{1/2}, \ldots, \lambda_n{}^{1/2}$, with $\mathbf{e}_1, \mathbf{e}_2, \ldots, \mathbf{e}_n$ as the corresponding eigenvectors. By the spectral theorem applied to the symmetric tensor **B**, it then follows that

$$\mathbf{B} = \lambda_1^{1/2}\,\mathbf{e}_1\otimes\mathbf{e}_1 + \lambda_2^{1/2}\,\mathbf{e}_2\otimes\mathbf{e}_2 + \ldots + \lambda_n^{1/2}\,\mathbf{e}_n\otimes\mathbf{e}_n. \tag{3.14}$$

So if there is a positive definite, symmetric tensor **B** such that $\mathbf{B}^2 = \mathbf{A}$, **B** must be given by (3.14). It remains only to verify that, if **B** is defined by (3.14), then **B** is symmetric, positive definite, and such that $\mathbf{B}^2 = \mathbf{A}$. The first two of these requirements are obviously satisfied. To establish the fact that $\mathbf{B}^2 = \mathbf{A}$, it is helpful to use the result of Problem 3.4, which implies that, if $\mathbf{e}_1, \mathbf{e}_2, \ldots, \mathbf{e}_n$ is an orthonormal set, then the tensor $\mathbf{P} = (\mathbf{e}_i\otimes\mathbf{e}_i)(\mathbf{e}_j\otimes\mathbf{e}_j)$ (no sum on i or j) is the null tensor if $i \neq j$, but coincides with $\mathbf{e}_i\otimes\mathbf{e}_i$ when $i = j$. Using this result to simplify the formula for $\mathbf{B}^2$ that follows from (3.14), we obtain $\mathbf{B}^2 = \lambda_1\,\mathbf{e}_1\otimes\mathbf{e}_1 + \lambda_2\,\mathbf{e}_2\otimes\mathbf{e}_2 + \ldots + \lambda_n\,\mathbf{e}_n\otimes\mathbf{e}_n$; according to (3.13), this is precisely **A**.

The geometrical effect of a symmetric tensor **A** on a vector **x** is easily appreciated if **x** is resolved into components in a principal basis $\mathbf{e}_1, \mathbf{e}_2, \ldots, \mathbf{e}_n$ for **A**: $\mathbf{x} = \xi_i\mathbf{e}_i$. Then as noted earlier, $\mathbf{Ax} = \Sigma\,\lambda_i\xi_i\,\mathbf{e}_i$, where the sum extends over the range of i. If $\mathbf{y} = \mathbf{Ax}$, then the k$^{th}$ component $\eta_k$ of **y** in the basis **e** is given by $\eta_k = \lambda_k\xi_k$ (no sum on k). Thus the magnitude $|\eta_k|$ of the k$^{th}$ component of **y** is *stretched* by a factor $|\lambda_k|$ if $|\lambda_k| > 1$, *contracted* by this factor if $|\lambda_k| < 1$, and annihilated if $\lambda_k = 0$. If $\lambda_k < 0$, then the sense of the contribution $\eta_k\,\mathbf{e}_k$ (no sum) to **y** is *opposite* to that of the associated contribution $\xi_k\,\mathbf{e}_k$ (no sum) to **x**, corresponding to a "reflection" in the direction $\mathbf{e}_k$, in addition to the stretch or contraction. If **A** is positive definite, then the reflections are absent, and the geometric action of **A** is that of a stretch (or contraction) of amount $\lambda_i$ in the direction of $\mathbf{e}_i$, $i = 1, \ldots, n$.

***Example 3.3.*** *A transformation on the space of trigonometric polynomials.* Let $\mathsf{F}_n$ be the n-dimensional real Euclidean space of trigonometric polynomials **p**(t) introduced in Example 1.8. Let

$$\mathbf{p}(t) = \alpha_0 + \sum_{k=1}^{N} (\alpha_k \cos kt + \beta_k \sin kt), \tag{3.15}$$

be a typical element of $F_n$, and let **A** be the linear transformation that carries **p** to its second derivative:

$$\mathbf{Ap} = \mathbf{p}'', \tag{3.16}$$

where the primes indicate differentiation with respect to t. Using the scalar product defined in (1.28), one finds that, for any two trigonometric polynomials **p**, **q** in $F_n$,

$$(\mathbf{p}, \mathbf{Aq}) - (\mathbf{Ap}, \mathbf{q}) = \int_0^{2\pi} [\mathbf{p}(t)\, \mathbf{q}''(t) - \mathbf{p}''(t)\mathbf{q}(t)]\, dt. \tag{3.17}$$

Integrating by parts and exploiting the fact that both **p** and **q** are periodic functions with period $2\pi$, one finds from (3.17) that $(\mathbf{p}, \mathbf{Aq}) - (\mathbf{Ap}, \mathbf{q}) = 0$, so that **A** is symmetric. What are its eigenvalues and eigenvectors? Suppose **p** is a trigonometric polynomial such that $\mathbf{Ap} = \lambda\mathbf{p}$ for some real number $\lambda$; then by (3.15) and (3.16),

$$\sum_{k=1}^{N} (-k^2 \alpha_k \cos kt - k^2 \beta_k \sin kt)$$
$$= \lambda \alpha_0 + \sum_{k=1}^{N} (\lambda \alpha_k \cos kt + \lambda \beta_k \sin kt). \tag{3.18}$$

It follows that

$$-\lambda \alpha_0 = 0 \quad \text{and} \quad -(\lambda + k^2)\, \alpha_k = 0,\ -(\lambda + k^2)\, \beta_k = 0$$
$$\text{for } k = 1, \ldots, N. \tag{3.19}$$

These represent $n = 2N + 1$ linear homogeneous equations for the n unknown $\alpha_k$s and $\beta_k$s. The determinant of this system, which is the characteristic polynomial of **A**, is given by

$$P_{\mathbf{A}}(\lambda) = -\lambda\, (\lambda + 1)^2(\lambda + 4)^2 \ldots (\lambda + j^2)^2 \ldots (\lambda + N^2)^2. \tag{3.20}$$

The eigenvalues of **A** are thus $\lambda_1 = 0$, $\lambda_2 = \lambda_3 = -1$, $\lambda_4 = \lambda_5 = -4$, ..., $\lambda_{2j} = \lambda_{2j+1} = -j^2$, ..., $\lambda_{n-1} = \lambda_n = -N^2$, where $n = 2N + 1$. The eigenvector corresponding to the eigenvalue $\lambda = 0$ is the constant trigonometric polynomial $\mathbf{p} = \alpha_0$. Each of the other eigenvalues has multiplicity two as a zero of $P_A(\lambda)$, and has two linearly independent corresponding eigenvectors: thus for $\lambda = \lambda_{2j} = \lambda_{2j+1}$, the associated eigenvectors are constant multiples of either $\mathbf{p} = \cos jt$ or $\mathbf{p} = \sin jt$. When these eigenvectors are scaled to have unit length, they coincide precisely with the vectors in the orthonormal basis for $\mathsf{F}_n$ given in (1.29).

***Example 3.4.*** *Symmetric tensors in a two-dimensional space.* Consider a two-dimensional real Euclidean space R. Let $\mathbf{A}$ be a symmetric, positive definite tensor, and suppose that $\mathbf{e}_1$, $\mathbf{e}_2$ are eigenvectors of $\mathbf{A}$ corresponding to positive eigenvalues $\lambda_1$, $\lambda_2$:

$$\mathbf{A}\mathbf{e}_i = \lambda_i \mathbf{e}_i \quad \text{(no sum on i)}. \tag{3.21}$$

We represent any vector $\mathbf{x}$ in R as $\mathbf{x} = \xi_j\, \mathbf{e}_j$, and we identify the components $\xi_1$, $\xi_2$ of $\mathbf{x}$ in this principal basis for $\mathbf{A}$ as the *coordinates* of a point whose position vector is $\mathbf{x}$. We view $\mathbf{A}$ as a mapping that carries the point $\mathbf{x}$ to the point $\mathbf{y} = \mathbf{A}\mathbf{x}$. Consider all the points $\mathbf{x}$ that lie on the circle $|\mathbf{x}| = 1$ of unit radius; where do these points go under the mapping $\mathbf{A}$? As shown in Problem 3.11, the images $\mathbf{y}$ of all such points $\mathbf{x}$ lie on an ellipse whose principal axes lie along the vectors $\mathbf{e}_1$ and $\mathbf{e}_2$; the semi-axes of the ellipse are of length $\lambda_1$ and $\lambda_2$. A similar geometrical discussion applies in three-dimensional spaces, and the analytical conclusions have clear counterparts in *any* finite dimensional real Euclidean space.

Some properties of skew-symmetric tensors are explored in the problems at the end of this chapter.

## Orthogonal Tensors

A second class of tensors of great practical importance is that of the *orthogonal* tensors. A linear transformation $\mathbf{Q}$ on a finite dimensional real Euclidean space R is said to be *orthogonal* if it preserves the length of every input vector $\mathbf{x}$: thus $|\mathbf{Q}\mathbf{x}| = |\mathbf{x}|$ for all $\mathbf{x}$ in R. We first observe that an orthogonal tensor $\mathbf{Q}$ also preserves the scalar product of any two vectors $\mathbf{x}$ and $\mathbf{y}$ in R, and therefore the angle between them. To see this, note first that

$$|\mathbf{Q}(\mathbf{x}-\mathbf{y})|^2 = (\mathbf{Q}(\mathbf{x}-\mathbf{y}), \mathbf{Q}(\mathbf{x}-\mathbf{y})) = (\mathbf{Q}\mathbf{x}, \mathbf{Q}\mathbf{x}) - 2(\mathbf{Q}\mathbf{x}, \mathbf{Q}\mathbf{y}) + (\mathbf{Q}\mathbf{y}, \mathbf{Q}\mathbf{y}) \quad \text{for all } \mathbf{x}, \mathbf{y} \text{ in R}. \tag{3.22}$$

But $|\mathbf{Q}(\mathbf{x}-\mathbf{y})| = |\mathbf{x}-\mathbf{y}|$, and $(\mathbf{Q}\mathbf{x}, \mathbf{Q}\mathbf{x}) = |\mathbf{Q}\mathbf{x}|^2 = |\mathbf{x}|^2$, $(\mathbf{Q}\mathbf{y}, \mathbf{Q}\mathbf{y}) = |\mathbf{Q}\mathbf{y}|^2 = |\mathbf{y}|^2$, so (3.22) becomes

$$|\mathbf{x}-\mathbf{y}|^2 = |\mathbf{x}|^2 + |\mathbf{y}|^2 - 2(\mathbf{Q}\mathbf{x}, \mathbf{Q}\mathbf{y}). \tag{3.23}$$

Expanding the left side of (3.23) leads immediately to

$$(\mathbf{x}, \mathbf{y}) = (\mathbf{Qx}, \mathbf{Qy}) \text{ for } \mathbf{x}, \mathbf{y} \text{ in } \mathsf{R}, \tag{3.24}$$

so that the scalar product is indeed preserved under the action of any orthogonal tensor **Q**.

We next show that every orthogonal tensor **Q** is non-singular, and that its inverse coincides with its transpose. The first of these assertions is virtually immediate, since $\mathbf{Qx} = \mathbf{o}$ implies that $|\mathbf{x}| = |\mathbf{o}| = 0$; the only vector of zero length is the null vector, so $\mathbf{Qx} = \mathbf{o}$ implies that $\mathbf{x} = \mathbf{o}$. Thus $N(\mathbf{Q}) = \{\mathbf{o}\}$, so by the finite dimensionality of R, **Q** is non-singular and thus has an inverse $\mathbf{Q}^{-1}$. Next, (3.24) can be written $(\mathbf{x}, \mathbf{y}) = (\mathbf{Q}^T\mathbf{Qx}, \mathbf{y})$; if **x** is momentarily fixed but arbitrary, this statement says that $\mathbf{x} - \mathbf{Q}^T\mathbf{Qx}$ is orthogonal to every vector **y** in R, whence $\mathbf{Q}^T\mathbf{Qx} = \mathbf{x}$ for every **x**. Thus $\mathbf{Q}^T\mathbf{Q} = \mathbf{1}$, and therefore $\mathbf{Q}^T\mathbf{Q}\mathbf{Q}^{-1} = \mathbf{Q}^{-1}$, or $\mathbf{Q}^T = \mathbf{Q}^{-1}$, as announced. But then

$$\mathbf{QQ}^T = \mathbf{Q}^T\mathbf{Q} = \mathbf{1}. \tag{3.25}$$

It is not hard to show that the conditions (3.25) are not only necessary for the orthogonality of **Q**, they are also sufficient; see Problem 3.18.

Let $\mathbf{e}_1, ..., \mathbf{e}_n$ be an orthonormal basis, and let $q_{ij}$ be the components of **Q** in **e**. Since the components of $\mathbf{Q}^T$ in **e** are $q_{ji}$, (3.25) requires that

$$q_{ik}\, q_{jk} = q_{ki}\, q_{kj} = \delta_{ij}, \tag{3.26}$$

so that the matrix $\underset{\sim}{Q}$ of components of **Q** in **e** is orthogonal in the sense of matrices.

Since according to Appendix 1, (a) the determinant of a product of tensors is the product of their determinants, (b) the determinants of a tensor and its transpose coincide, and (c) the determinant of **1** is unity, it follows from (3.25) that

$$\det \mathbf{Q} = \pm 1. \tag{3.27}$$

We say that **Q** is a *proper* orthogonal tensor if $\det \mathbf{Q} = 1$, *improper* if $\det \mathbf{Q} = -1$. We interpret this distinction below.

What can be said about the eigenvalues of an orthogonal tensor **Q**? If **Q** *has* an eigenvalue $\lambda$, and if **f** is a corresponding eigenvector, then $\mathbf{Qf} = \lambda\mathbf{f}$. But $|\mathbf{Qf}| = |\mathbf{f}|$, so any eigenvalue of **Q** necessarily satisfies $|\lambda| = 1$. It follows that the zeros of the characteristic polynomial $P_Q(\lambda)$ of **Q** are one, minus one, or complex; if there are complex zeros, they of course occur in conjugate pairs. Suppose **Q** has $\lambda = 1$ as an eigenvalue; then any corresponding eigenvector **f** satisfies $\mathbf{Qf} = \mathbf{f}$, and the invariant one-dimensional linear manifold for **Q** that corresponds to this eigenvalue consists entirely of vectors that are left unchanged by the action of **Q**. Similarly, if $\lambda = -1$ is an eigenvalue of **Q**, every vector in an associated one-dimensional invariant linear manifold is *reflected* by **Q**. It is easily seen that, if $\lambda$ is an eigenvalue of **Q**, it is also an eigenvalue of the inverse $\mathbf{Q}^T$ of **Q**, and every corresponding eigenvector of **Q** is also an eigenvector of $\mathbf{Q}^T$ for this eigenvalue. It follows that every one-dimensional invariant linear manifold for **Q** is also invariant for $\mathbf{Q}^T$, and *vice versa*.

If $\lambda = \mu + i\nu$ is a *complex* zero of $P_Q(\lambda)$, then there are linearly independent vectors **f**, **g** generating a two-dimensional invariant linear manifold for **Q** and satisfying

$$\mathbf{Qf} = \mu\mathbf{f} - \nu\mathbf{g}, \quad \mathbf{Qg} = \nu\mathbf{f} + \mu\mathbf{g}. \tag{3.28}$$

The fact that $|\mathbf{f}|$, $|\mathbf{g}|$ and $(\mathbf{f}, \mathbf{g})$ are all left unchanged when **f** and **g** are replaced by **Qf** and **Qg** leads from (3.28) to the following three equations that must be satisfied by $|\mathbf{f}|^2$, $|\mathbf{g}|^2$ and $(\mathbf{f}, \mathbf{g})$:

$$\left.\begin{aligned} (\mu^2 - 1)\,|\mathbf{f}|^2 + \nu^2\,|\mathbf{g}|^2 - 2\mu\nu\,(\mathbf{f}, \mathbf{g}) &= 0, \\ \nu^2\,|\mathbf{f}|^2 + (\mu^2 - 1)\,|\mathbf{g}|^2 + 2\mu\nu\,(\mathbf{f}, \mathbf{g}) &= 0, \\ \mu\nu\,|\mathbf{f}|^2 - \mu\nu\,|\mathbf{g}|^2 + (\mu^2 - \nu^2 - 1)\,(\mathbf{f}, \mathbf{g}) &= 0. \end{aligned}\right\} \tag{3.29}$$

It is easy to show that these equations imply that $\mu^2 + \nu^2 = 1$, and that $|\mathbf{f}| = |\mathbf{g}|$ as well as $(\mathbf{f}, \mathbf{g}) = 0$. Thus **f** and **g** are orthogonal, and since $|\mathbf{f}|$ and $|\mathbf{g}|$ have the same length, (3.28) shows that we may without loss of generality assume that **f** and **g** are unit vectors; otherwise, we could divide both **f** and **g** by their common length, producing unit vectors that still satisfy (3.28). Moreover, the fact that $\mu^2 + \nu^2 = 1$ means that there is a real angle $\varphi$ such that $\mu = \cos\varphi$, $\nu = \sin\varphi$, so that the complex zero $\lambda = \mu + i\nu$ of $P_Q(\lambda)$ is given by $\lambda = e^{i\varphi}$. (One may note in passing that all zeros of $P_A(\lambda)$ lie on the unit circle in the complex plane.) The two-dimensional invariant linear manifold for **Q** that corresponds to this $\lambda$ is thus generated by the orthonormal pair **f**, **g** satisfying (3.28), which may now be written as

$$\mathbf{Qf} = \cos\varphi\,\mathbf{f} - \sin\varphi\,\mathbf{g}, \quad \mathbf{Qg} = \sin\varphi\,\mathbf{f} + \cos\varphi\,\mathbf{g} \tag{3.30}$$

It follows that *every* vector in the two-dimensional invariant linear manifold corresponding to the zero $\lambda = e^{i\varphi}$, when acted upon by **Q**, is rotated within the manifold through an angle $\varphi$.

The result of Problem 3.19 shows that a two-dimensional invariant linear manifold for the orthogonal tensor **Q** is also a two-dimensional invariant linear manifold for $\mathbf{Q}^T$.

We can now determine the form of the most general orthogonal tensor **Q** on an n-dimensional real Euclidean space. First, let $\mathsf{K}$ be the set of all vectors **x**, including the null vector, satisfying $\mathbf{Qx} = \mathbf{x}$; all but $\mathbf{x} = \mathbf{o}$ of these are thus eigenvectors corresponding to the eigenvalue $\lambda = 1$ for **Q**. Clearly $\mathsf{K}$ is an invariant linear manifold for **Q**; although $\mathsf{K}$ *might* contain *only* the null vector, we assume temporarily that this is not the case, so that the dimension k of K is at least one. Similarly, let H be the set of all vectors $\mathbf{y}$ such that $\mathbf{Qy} = -\mathbf{y}$, including $\mathbf{y} = \mathbf{o}$; H is also an invariant linear manifold for $\mathbf{Q}$, and we temporarily assume that its dimension h is also positive. If $\mathbf{x}$ is in K and $\mathbf{y}$ is in H, one has $(\mathbf{x}, \mathbf{y}) = (\mathbf{Qx}, \mathbf{Qy}) = (\mathbf{x}, -\mathbf{y}) = -(\mathbf{x}, \mathbf{y})$, so that $(\mathbf{x}, \mathbf{y}) = 0$. Thus the two manifolds K and H are "perpendicular," in the sense that every vector in one manifold is orthogonal to all vectors in the other. Let $\mathbf{k}_1, ..., \mathbf{k}_k$ be an orthonormal basis for K, $\mathbf{h}_1, ..., \mathbf{h}_h$ an orthonormal basis for H. Now introduce the linear manifold M = span $\{\mathbf{k}_1, ..., \mathbf{k}_k, \mathbf{h}_1 ..., \mathbf{h}_h\}$; the $\mathbf{k}_i$s and $\mathbf{h}_i$s comprise an orthonormal basis for M, whose dimension is clearly k + h. It is easy to verify that M is also an *invariant* linear manifold, not only for $\mathbf{Q}$, but for $\mathbf{Q}^T$ as well. Set $M^\perp = \{\mathbf{z} \text{ in } R \mid (\mathbf{z}, \mathbf{w}) = 0 \text{ for every } \mathbf{w} \text{ in } M\}$, and let $\mathbf{z}$ be any vector in $M^\perp$. Then for any $\mathbf{w}$ in M, $(\mathbf{Qz}, \mathbf{w}) = (\mathbf{z}, \mathbf{Q}^T\mathbf{w})$; but $\mathbf{Q}^T\mathbf{w}$ is in M when $\mathbf{w}$ is in M, so $(\mathbf{z}, \mathbf{Q}^T\mathbf{w}) = 0$. Thus whenever $\mathbf{z}$ is in $M^\perp$, so is $\mathbf{Qz}$, making $M^\perp$ *also* an invariant linear manifold for $\mathbf{Q}$. Conceivably, $M^\perp$ contains only the null vector; we assume temporarily that this is *not* the case, so that the dimension of $M^\perp$ is positive.

Now consider $\mathbf{Q}$ as an orthogonal linear transformation restricted to the smaller real Euclidean space $M^\perp$. There can be no non-null vectors $\mathbf{z}$ in $M^\perp$ such that $\mathbf{Qz} = \pm\mathbf{z}$, since all such vectors are in M. Thus the characteristic polynomial of $\mathbf{Q}$, when the latter is restricted to $M^\perp$, can have neither $\lambda = 1$ nor $\lambda = -1$ as a zero, and hence all of its zeros must be complex; the dimension of $M^\perp$ must therefore be even, say 2m, where m is a positive integer. Moreover, the restriction of $\mathbf{Q}$ to $M^\perp$ must have at least one two-dimensional invariant linear manifold, say $M_1$, and there must be an orthonormal pair of vectors $\mathbf{u}_1$, $\mathbf{v}_1$ in $M_1$ and an angle $\varphi_1$ such that

$$\mathbf{Qu}_1 = \cos\varphi_1\,\mathbf{u}_1 - \sin\varphi_1\,\mathbf{v}_1, \quad \mathbf{Qv}_1 = \sin\varphi_1\,\mathbf{u}_1 + \cos\varphi_1\,\mathbf{v}_1. \qquad (3.31)$$

Next, consider the *orthogonal complement* $M_1^\perp/M^\perp$ *of* $M_1$ *relative to* $M^\perp$, defined by

$$M_1^\perp/M^\perp = \{\mathbf{z} \text{ in } M^\perp \mid (\mathbf{z}, \mathbf{w}) = 0 \text{ for every } \mathbf{w} \text{ in } M_1\}. \qquad (3.32)$$

We now show that this relative orthogonal complement is *also* an invariant linear manifold for $\mathbf{Q}$. Let $\mathbf{z}$ be in $M_1^\perp/M^\perp$, and let $\mathbf{w}$ be in $M_1$. Because $M_1$ is an invariant linear manifold for $\mathbf{Q}^T$ as well as for $\mathbf{Q}$, $\mathbf{z}$ and $\mathbf{Q}^T\mathbf{w}$ will be orthogonal: $(\mathbf{z}, \mathbf{Q}^T\mathbf{w}) = 0$. But then $(\mathbf{Qz}, \mathbf{w}) = 0$ as well, so that $\mathbf{Qz}$ is orthogonal to every $\mathbf{w}$ in $M_1$. Hence $\mathbf{Qz}$ is in $M_1^\perp/M^\perp$ whenever $\mathbf{z}$ is, so $M_1^\perp/M^\perp$ is an invariant linear manifold for $\mathbf{Q}$. Assuming for the moment that this manifold contains more than just the null vector, it must contain a two-dimensional invariant linear manifold $M_2$ for $\mathbf{Q}$, and there must be generating orthonormal vectors $\mathbf{u}_2$, $\mathbf{v}_2$ in $M_2$ and an angle $\varphi_2$ such that

$$\mathbf{Q}\mathbf{u}_2 = \cos\varphi_2\,\mathbf{u}_2 - \sin\varphi_2\,\mathbf{v}_2, \quad \mathbf{Q}\mathbf{v}_2 = \sin\varphi_2\,\mathbf{u}_2 + \cos\varphi_2\,\mathbf{v}_2. \quad (3.33)$$

We continue in this way until $\mathsf{M}^\perp$ is exhausted, generating m two-dimensional invariant linear manifolds $\mathsf{M}_1, \mathsf{M}_2, ..., \mathsf{M}_m$ for $\mathbf{Q}$, each lying within the orthogonal complement $\mathsf{M}^\perp$ of the span $\mathsf{M}$ of the eigenvectors of $\mathbf{Q}$. Moreover, the vectors $\mathbf{u}_j$, $\mathbf{v}_j$ form an orthonormal basis for $\mathsf{M}_j$, $j = 1, ..., m$, and they satisfy the pair of equations obtained from (3.33) by replacing $\mathbf{u}_2$, $\mathbf{v}_2$ by $\mathbf{u}_j$, $\mathbf{v}_j$ and $\varphi_2$ by a suitable angle $\varphi_j$.

By construction, it is clear that the vectors $\mathbf{k}_1, ..., \mathbf{k}_k, \mathbf{h}_1, ..., \mathbf{h}_h, \mathbf{u}_1, \mathbf{v}_1, ..., \mathbf{u}_m, \mathbf{v}_m$ constitute an orthonormal basis for the entire space $\mathsf{R}$, and that $k + h + 2m = n$, the dimension of $\mathsf{R}$. Because the $\mathbf{k}_j$s and $\mathbf{h}_j$s satisfy $\mathbf{Q}\mathbf{k}_j = \mathbf{k}_j$ and $\mathbf{Q}\mathbf{h}_j = -\mathbf{h}_j$, respectively, and in view of the equation satisfied by the pairs $\mathbf{u}_j$, $\mathbf{v}_j$, one can immediately determine the matrix $\underline{Q}$ of components of the orthogonal tensor $\mathbf{Q}$ in this special basis. This matrix is best described in terms of smaller block matrices $\underline{K}$, $\underline{H}$, and $\underline{M}_1, \underline{M}_2, ..., \underline{M}_m$ that correspond to the action of $\mathbf{Q}$ on the invariant linear manifolds $\mathsf{K}$, $\mathsf{H}$, and $\mathsf{M}_1, \mathsf{M}_2, ..., \mathsf{M}_m$, respectively; K is the $k \times k$ identity matrix $\underline{1}_k$, $\underline{H} = -\underline{1}_h$ is the negative of the $h \times h$ identity matrix, and $\underline{M}_j$ is the $2 \times 2$ "rotation matrix" given by

$$\underline{M}_j = \begin{pmatrix} \cos\varphi_j & -\sin\varphi_j \\ \sin\varphi_j & \cos\varphi_j \end{pmatrix}, \; j = 1, ..., m. \quad (3.34)$$

The submatrices $\underline{K}$, $\underline{H}$, and $\underline{M}_1, ..., \underline{M}_m$ are to be embedded in $\underline{Q}$ so that their principal diagonals lie along that of $\underline{Q}$, with the identity block $\underline{K}$ in the upper left $k \times k$ corner, the reflection block $\underline{H}$ occupying the next $h \times h$ submatrix along the diagonal, followed successively by the $2 \times 2$ rotation blocks $\underline{M}_1, ..., \underline{M}_m$, also hung along the principal diagonal. We have therefore established the following proposition:

---

***Proposition 3.5.*** Let $\mathbf{Q}$ be an orthogonal tensor. Then there is an orthonormal basis in which the matrix $\underline{Q}$ of $\mathbf{Q}$ is given by

---

$$\underline{Q} = \text{block diag}\,(\underline{K}, \underline{H}, \underline{M}_1, ..., \underline{M}_m). \quad (3.35)$$

We view (3.35) as the canonical matrix representation of an orthogonal tensor in an n-dimensional real Euclidean space $\mathsf{R}$.

In deriving (3.35), we assumed that the respective dimensions k, h, and 2m of $\mathsf{K}$, $\mathsf{H}$, and $\mathsf{M}^\perp$ were all positive. If either k or h vanishes, then the corresponding block is simply absent in the representation (3.35); if m vanishes, all of the blocks $\underline{M}_j$ are absent.

From (3.35) and the properties of determinants, we can immediately infer that

$$\det \mathbf{Q} = (-1)^h \det \underline{M}_1 \det \underline{M}_2 \ldots \det \underline{M}_m = (-1)^h. \tag{3.36}$$

Thus $\mathbf{Q}$ is proper if the number h of reflections is even, improper otherwise. If $\mathbf{Q}$ is proper, so that h is even, say $h = 2p$ for a positive integer p, then the $h \times h$ block $\underline{H}$ in the matrix $\underline{Q}$ can be replaced by p blocks of the form (3.34) in each of which the angle $\varphi = \pi$, corresponding to rotation through an angle $\pi$ in the two-dimensional linear manifold spanned by two eigenvectors of $\mathbf{Q}$ corresponding to the eigenvalue $-1$. Thus one loses no generality by assuming that the $\underline{H}$-block is *absent* if $\mathbf{Q}$ is proper, while $\underline{H}$ is the "$1 \times 1$" matrix $-1$ if $\mathbf{Q}$ is improper. A proper orthogonal tensor is therefore formed exclusively of rotations in two-dimensional invariant linear manifolds, possibly together with lower-dimensional identity transformations. An improper orthogonal tensor consists of rotations, identities, and a reflection. We shall speak of a proper orthogonal tensor as a *rotation tensor.*

***Example 3.5.*** *Three-dimensional real Euclidean spaces.* Suppose the real Euclidean space under consideration has dimension three, and let $\mathbf{Q}$ be an orthogonal tensor. It is not difficult to see from the discussion leading to (3.35) that there must be an orthonormal basis in which the matrix of components $\underline{Q}$ of $\mathbf{Q}$ has one of the following two forms:

$$\underline{Q} = \begin{pmatrix} 1 & 0 & 0 \\ 0 & \cos\varphi & -\sin\varphi \\ 0 & \sin\varphi & \cos\varphi \end{pmatrix} \quad \text{or } \underline{Q} = \begin{pmatrix} -1 & 0 & 0 \\ 0 & \cos\varphi & -\sin\varphi \\ 0 & \sin\varphi & \cos\varphi \end{pmatrix}. \tag{3.37}$$

The first of these representations arises if $\mathbf{Q}$ is proper, the second if $\mathbf{Q}$ is improper. Observe that the case in which $\mathbf{Q}$ coincides with the identity tensor $\mathbf{1}$ corresponds to the first representation with $\varphi = 0$, while $\mathbf{Q} = -\mathbf{1}$ leads to the second matrix with $\varphi = \pi$. The first alternative in (3.37) describes a pure rotation about the 1-axis through an angle $\varphi$. The second matrix in (3.37) corresponds to the same rotation, together with a reflection in the 1-axis.

In the preceding chapter, we derived the change-of basis-formula (2.37) for the components of a linear transformation on any finite dimensional vector space. We return now to this matter in a real *Euclidean* space $\mathsf{R}$, assuming that the two bases $\mathbf{e}_1, \ldots, \mathbf{e}_n$ and $\mathbf{f}_1, \ldots, \mathbf{f}_n$ under consideration are both orthonormal. Define a tensor $\mathbf{R}$ by specifying its action on the basis $\mathbf{e}$ as follows:

$$\mathbf{R}\mathbf{e}_i = \mathbf{f}_i \tag{3.38}$$

Since the basis $\mathbf{f}$ is orthonormal, $(\mathbf{R}\mathbf{e}_i, \mathbf{R}\mathbf{e}_j) = \delta_{ij}$. As a result, if $\mathbf{x} = \xi_i \mathbf{e}_i$ is any vector in $\mathsf{R}$, we have $|\mathbf{R}\mathbf{x}|^2 = (\mathbf{R}\mathbf{x}, \mathbf{R}\mathbf{x}) = \xi_i \xi_j (\mathbf{R}\mathbf{e}_i, \mathbf{R}\mathbf{e}_j) = \delta_{ij}\, \xi_i \xi_j = \xi_i \xi_i =$

$|\mathbf{x}|^2$, so the tensor **R** preserves length and is therefore orthogonal. Now suppose that **A** is any tensor; its components $a^{\mathbf{f}}_{ij}$ in the basis **f** are given by

$$a^{\mathbf{f}}_{ij} = (\mathbf{f}_i, \mathbf{A}\mathbf{f}_j) = (\mathbf{R}\mathbf{e}_i, \mathbf{A}\mathbf{R}\mathbf{e}_j) = (\mathbf{e}_i, \mathbf{R}^T\mathbf{A}\mathbf{R}\mathbf{e}_j). \tag{3.39}$$

The last entries on the right in (3.39) are the components in **e** of the tensor $\mathbf{R}^T\mathbf{A}\mathbf{R}$; thus from (3.39) we infer the change-of-orthonormal-basis formula in the alternative forms

$$a^{\mathbf{f}}_{ij} = r^{\mathbf{e}}_{ki}\, a^{\mathbf{e}}_{kl}\, r^{\mathbf{e}}_{lj}, \quad \underline{A}^{\mathbf{f}} = (\underline{R}^{\mathbf{e}})^T \underline{A}^{\mathbf{e}} \underline{R}^{\mathbf{e}}, \tag{3.40}$$

where $r^{\mathbf{e}}_{ij}$ are the components of **R** in **e**, and $\underline{R}^{\mathbf{e}}$ is the matrix of components of **R** in **e**. Alternatively, we could have derived (3.40) directly from (2.37).

For later purposes, it is useful to describe here the *additive decomposition* of an arbitrary tensor **A** into symmetric and skew-symmetric parts. Let $\mathbf{S} = (1/2)(\mathbf{A} + \mathbf{A}^T)$, $\boldsymbol{\Omega} = (1/2)(\mathbf{A} - \mathbf{A}^T)$; since the transpose of a linear combination of tensors is the corresponding linear combination of their separate transposes, it is clear that **S** is a symmetric tensor, $\boldsymbol{\Omega}$ a skew-symmetric one. Moreover, $\mathbf{A} = \mathbf{S} + \boldsymbol{\Omega}$, so that any **A** can be represented as the sum of a symmetric tensor and a skew-symmetric one. It is easy to show that **S** and $\boldsymbol{\Omega}$ as defined above are the *only* symmetric and skew-symmetric tensors, respectively, for which $\mathbf{A} = \mathbf{S} + \boldsymbol{\Omega}$, so that the additive decomposition of a tensor into its symmetric and skew-symmetric parts is unique.

According to the decomposition $\mathbf{A} = \mathbf{S} + \boldsymbol{\Omega}$, the components of **A**, **S**, and $\boldsymbol{\Omega}$ in any orthonormal basis **e** are related by $a^{\mathbf{e}}_{ij} = s^{\mathbf{e}}_{ij} + \omega^{\mathbf{e}}_{ij}$. If we choose for **e** a *principal* basis for **S**, then $s^{\mathbf{e}}_{ii} = \lambda_i$ (no sum), where the $\lambda_i$ are the eigenvalues of **S**, and of course $\omega_{ii} = 0$ (no sum). Thus $a^{\mathbf{e}}_{ii} = \lambda_i$ (no sum), so that

$$\text{Tr}\,\mathbf{A} = \sum_{i=1}^{n} a^{\mathbf{f}}_{ii} = \lambda_1 + \lambda_2 + \ldots + \lambda_n, \tag{3.41}$$

where **f** is *any* basis, and the $\lambda_i$s are the eigenvalues of the symmetric part **S** of **A**. As shown earlier, Tr **A** is a scalar invariant for **A**; in particular, it has the same values when computed in two orthonormal bases **e** and **f** related by $\mathbf{f} = \mathbf{Q}\mathbf{e}$, where **Q** is an orthogonal tensor. It is easy to show that this implies that

$$\text{Tr}\,(\mathbf{Q}^T\mathbf{A}\mathbf{Q}) = \text{Tr}\,(\mathbf{Q}\mathbf{A}\mathbf{Q}^T) = \text{Tr}\,\mathbf{A} \tag{3.42}$$

for any orthogonal **Q**, a fact that will be used later.

## Polar Decomposition of a Tensor

We turn now to a result that asserts that the geometric effect of *any* tensor may be understood in terms of the geometric properties of symmetric and orthogonal tensors. This proposition is vital in the description of the kinematics of solids undergoing deformation; some aspects of this application will be discussed Chapter 5.

---

***Proposition 3.6.*** Let **A** be a non-singular tensor. Then there are unique symmetric positive definite tensors **U**, **V** and a unique orthogonal tensor **Q** such that

$$\mathbf{A} = \mathbf{QU} = \mathbf{VQ}; \tag{3.43}$$

Moreover, $\mathbf{U} = \mathbf{Q}^T\mathbf{VQ}$.

---

The first of the representations of **A** in (3.43) is called the *right polar decomposition* of **A**, while the second is the *left* polar decomposition. These representations may be viewed as analogous to the "polar representation" $z = r\exp(i\theta)$ of a complex number $z = x + iy$; **A** plays the role of z, either **U** or **V** is the counterpart of r, and **Q** is the analog of $\exp(i\theta)$.

To demonstrate this proposition, we first derive conditions that are *necessary* for the *first* of the representations in (3.43) to hold. Indeed, if $(3.43)_1$ holds for some orthogonal tensor **Q** and some positive definite, symmetric tensor **U**, then $\mathbf{A}^T = \mathbf{UQ}^T$. It follows that $\mathbf{A}^T\mathbf{A} = \mathbf{U}^2$. Moreover, if **x** is any vector, $f(\mathbf{x}) = (\mathbf{x}, \mathbf{A}^T\mathbf{Ax}) = (\mathbf{Ax}, \mathbf{Ax}) = |\mathbf{Ax}|^2$; thus the quadratic form associated with the symmetric tensor $\mathbf{A}^T\mathbf{A}$ vanishes at **x** if an only if $|\mathbf{Ax}| = 0$, which is to say $\mathbf{Ax} = \mathbf{o}$. Since **A** is non-singular, $\mathbf{Ax} = \mathbf{o}$ if and only if $\mathbf{x} = \mathbf{o}$. Thus if $\mathbf{x} \neq \mathbf{o}$, the quadratic form $f(\mathbf{x})$ is positive, and therefore the symmetric tensor $\mathbf{A}^T\mathbf{A}$ is positive definite. It therefore has a unique symmetric, positive definite square root $(\mathbf{A}^T\mathbf{A})^{1/2}$; since **U** is symmetric and positive definite and satisfies $\mathbf{U}^2 = \mathbf{A}^T\mathbf{A}$, **U** must be given by $(\mathbf{A}^T\mathbf{A})^{1/2}$. Since **U** is positive definite, its eigenvalues must all be positive; it follows that **U** is non-singular. We then conclude that, if $(3.43)_1$ holds, $\mathbf{Q} = \mathbf{AU}^{-1}$. We have therefore shown that $(3.43)_1$, with **U** symmetric and positive definite and **Q** orthogonal, implies that $\mathbf{U} = (\mathbf{A}^T\mathbf{A})^{1/2}$ and $\mathbf{Q} = \mathbf{AU}^{-1}$.

Conversely, let **U** and **Q** be defined by the last two equations. Then clearly **U** is symmetric and positive definite, and $\mathbf{A} = \mathbf{QU}$. It remains only to show that **Q** is orthogonal. But $\mathbf{QQ}^T = \mathbf{AU}^{-1}\mathbf{U}^{-1}\mathbf{A}^T = \mathbf{A}(\mathbf{U}^2)^{-1}\mathbf{A} = \mathbf{A}(\mathbf{A}^T\mathbf{A})^{-1}\mathbf{A}^T = \mathbf{1}$ and $\mathbf{Q}^T\mathbf{Q} = \mathbf{U}^{-1}\mathbf{A}^T\mathbf{AU}^{-1} = \mathbf{U}^{-1}\mathbf{U}^2\mathbf{U}^{-1} = \mathbf{1}$, establishing the orthogonality of **Q** and completing the proof of the existence and uniqueness of the right polar decomposition $(3.43)_1$. An entirely analogous argument establishes a left polar decomposition $\mathbf{A} = \mathbf{VR}$, where **V** is symmetric and positive definite and **R** is orthogonal. From this it also follows that $\mathbf{A} = \mathbf{R}(\mathbf{R}^T\mathbf{VR})$. According to Problem 3.22, if **V** is symmetric and positive definite and $\mathbf{R}$ is orthogonal, then $\mathbf{R}^T\mathbf{V}\mathbf{R}$ is also symmetric and positive definite. Thus $\mathbf{A} = \mathbf{R}(\mathbf{R}^T\mathbf{V}\mathbf{R})$ is *also* a *right* polar decomposition of $\mathbf{A}$. But the uniqueness of the right decomposition implies that $\mathbf{R} = \mathbf{Q}$ and $\mathbf{R}^T\mathbf{V}\mathbf{R} = \mathbf{U}$. Thus both decompositions in (3.43) are established, as is the relation $\mathbf{U} = \mathbf{Q}^T\mathbf{V}\mathbf{Q}$.

According to the *right* polar decomposition, the geometric effect of any non-singular tensor may now be viewed as follows: first, there is a stretch (or contraction) of amount $\lambda_i$ in each of the n directions $\mathbf{e}_i$, where $\lambda_i$ and $\mathbf{e}_i$ are the eigenvalues and corresponding eigenvectors of the positive definite, symmetric "right stretch tensor" $\mathbf{U}$. These stretches are then followed by the rotations and reflections that make up the orthogonal factor $\mathbf{Q}$. Alternatively, the *left* decomposition says that we may view the rotations and reflections associated with $\mathbf{Q}$ as taking place first, followed by stretches whose magnitudes and directions are respectively those of the eigenvalues $\mu_i$ and eigenvectors $\mathbf{f}_i$ of the positive definite, left stretch tensor $\mathbf{V}$. As shown in Problem 3.22, one has $\mu_i = \lambda_i$, so that the magnitudes of the stretches associated with $\mathbf{U}$ are the same as those for $\mathbf{V}$, and the directions $\mathbf{e}_i$ and $\mathbf{f}_i$ for $\mathbf{U}$ and $\mathbf{V}$ are related by $\mathbf{f}_i = \mathbf{Q}\mathbf{e}_i$.

We shall study geometric implications of the polar decomposition theorem in greater detail in connection with the application to kinematics discussed in the final chapter.

Problem 3.28 is concerned with a version of the polar decomposition in which $\mathbf{A}$ is permitted to be singular. This more general result replaces (3.43) by the assertions that $\mathbf{A} = \mathbf{Q}\mathbf{U} = \mathbf{V}\mathbf{R}$, where $\mathbf{U}$, $\mathbf{V}$ are non-negative definite, uniquely determined symmetric tensors, while $\mathbf{Q}$ and $\mathbf{R}$ are orthogonal tensors that need not be unique and therefore need not be the same. The more restrictive version proved above is all that is needed in most applications.

## References

[3.1] R. Courant and D. Hilbert, *Methods of Mathematical Physics*, Volume 1, Interscience Press, New York, 1953.

[3.2] I.M. Gel'fand, *Lectures on Linear Algebra*, translated from the Russian by A. Shenitzer, Dover Publications, New York, 1989.

[3.3] P.R. Halmos, *Finite Dimensional Vector Spaces*, Second Edition, Van Nostrand-Reinhold, New York, 1958.

## Problems

**3.1.** A *linear function* on a real Euclidean space $\mathsf{R}$ is a real-valued function $\psi$ defined on $\mathsf{R}$ with the property that $\psi(\alpha\mathbf{x} + \beta\mathbf{y}) = \alpha\,\psi(\mathbf{x}) + \beta\,\psi(\mathbf{y})$ for every pair of real numbers $\alpha$, $\beta$ and every pair of vectors $\mathbf{x}$, $\mathbf{y}$ in

R. (a) Let **h** be a fixed vector in R. Show that $\psi(\mathbf{x}) = (\mathbf{h}, \mathbf{x})$ defines a linear function on R. (b) Let R be the real Euclidean space C of all continuous functions on $[0, \pi]$, the scalar product being that in (1.24). Show that the *average value* on $[0, \pi]$ of any function in C is an example of a linear function on R. (c) Suppose from here on that R is finite dimensional, and let $\mathbf{e}_1, ..., \mathbf{e}_n$ be an orthonormal basis for R. If $\psi$ is a linear function on R, let $h_j = \psi(\mathbf{e}_j)$ be the value of $\psi$ at $\mathbf{e}_j$. Set $\mathbf{h} = h_j \mathbf{e}_j$. If $\mathbf{x} = \xi_j \mathbf{e}_j$ is any vector in R, show that $\psi(\mathbf{x}) = \xi_j h_j = (\mathbf{h}, \mathbf{x})$. This shows that when R is finite dimensional, the scalar product is the *only* example of a linear function on R. For a given $\psi$, could there be more than one vector **h** such that $\psi(\mathbf{x}) = (\mathbf{h}, \mathbf{x})$?

**3.2.** Let **A** and **B** be linear transformations on a real Euclidean space R with respective transposes $\mathbf{A}^T$ and $\mathbf{B}^T$. Show that $\mathbf{C} = \mathbf{AB}$ has a transpose $\mathbf{C}^T$, and that it is given by $\mathbf{C}^T = \mathbf{B}^T\mathbf{A}^T$. Do not assume that R is finite dimensional.

**3.3.** Show that the tensor product $\mathbf{a}\otimes\mathbf{b}$ is a symmetric tensor if and only if one of the following three conditions holds: (a) $\mathbf{a} = \mathbf{o}$, (b) $\mathbf{b} = \mathbf{o}$, (c) $\mathbf{a} = k\mathbf{b}$, where k is a scalar.

**3.4.** Let **a** and **b** be vectors in a finite dimensional real Euclidean space, with $(\mathbf{a}, \mathbf{b}) = 0$ and $|\mathbf{a}| = 1$. Let $\mathbf{A} = \mathbf{a}\otimes\mathbf{a}$, $\mathbf{B} = \mathbf{b}\otimes\mathbf{b}$, $\mathbf{C} = \mathbf{AB}$, $\mathbf{D} = \mathbf{A}^2$. Show that $\mathbf{C} = \mathbf{O}$ and $\mathbf{D} = \mathbf{A}$.

**3.5.** Find the eigenvalues and eigenvectors of the tensor product $\mathbf{a}\otimes\mathbf{b}$ if $\mathbf{a} \neq \mathbf{o}$, $\mathbf{b} \neq \mathbf{o}$.

**3.6.** If **A** is a linear transformation of an n-dimensional linear vector space into itself, the *rank* of **A**, denoted by $r(\mathbf{A})$, is defined to be $r(\mathbf{A}) = n - \dim N(\mathbf{A})$, where $N(\mathbf{A})$ is the null space of **A**, and $\dim N(\mathbf{A})$ is the dimension of the null space. Let R be an n-dimensional real Euclidean space. (a) Let **a** and **b** be vectors in R, and set $\mathbf{A} = \mathbf{a}\otimes\mathbf{b}$. Show that either $r(\mathbf{A}) = 0$ or $r(\mathbf{A}) = 1$. (b) Let **A** be a Cartesian tensor on R whose rank is one. Show that there are vectors **a** and **b** with $\mathbf{a} \neq \mathbf{o}$ and $|\mathbf{b}| = 1$ such that $\mathbf{A} = \mathbf{a}\otimes\mathbf{b}$.

**3.7.** In an n-dimensional real Euclidean space R, two tensors **A** and **B** are said to be *rank-one-connected* if there are non-null vectors **a** and **b** such that $\mathbf{A} - \mathbf{B} = \mathbf{a}\otimes\mathbf{b}$. Show that if **A** is rank-one-connected to the identity tensor **1**, then $\lambda = 1$ is an eigenvalue of **A**. [Rank-one-connected tensors are of fundamental importance in continuum-mechanical theories of phase transformations in solids.]

**3.8.** Consider the real Euclidean space $R_2$, which is a special case of the space discussed in Example 3.1. Let $\mathbf{e}_1 = \binom{1}{0}$, $\mathbf{e}_2 = \binom{0}{1}$ and $\mathbf{f}_1 = \binom{1}{0}$, $\mathbf{f}_2 =$

$\binom{1}{1}$ be two bases for R. Note that **e** is orthonormal, but **f** is not. Let **A** be a special case of the linear transformation in (3.5) for which $n = 2$ and $a_{11} = a_{22} = 0$, $a_{12} = a_{21} = 1$. What are the components $a^{\mathbf{e}}_{ij}$ and $a^{\mathbf{f}}_{ij}$ of **A** in the respective bases **e** and **f**? Observe that $a^{\mathbf{f}}_{12} \neq a^{\mathbf{f}}_{21}$. Is **A** symmetric?

**3.9.** Show that the quadratic form $f(\mathbf{x}) = (\mathbf{Ax}, \mathbf{x})$ associated with a symmetric tensor **A** is positive for all non-null vectors **x** if and only if **A** is positive definite; we then speak of $f(\mathbf{x})$ as a *positive definite quadratic form.*

**3.10.** Let **A** be a skew-symmetric tensor. Suppose that **x** satisfies $\mathbf{Ax} = \lambda\mathbf{x}$ for some scalar $\lambda$. Show that $\lambda\,(\mathbf{x}, \mathbf{x}) = (\mathbf{x}, \mathbf{Ax}) = -\lambda\,(\mathbf{x}, \mathbf{x})$, so that either $\mathbf{x} = \mathbf{o}$ or $\lambda = 0$. Thus show that if a skew-symmetric tensor has an eigenvalue, the eigenvalue must be zero, and **A** must be singular. Give examples to show that a skew-symmetric tensor may or may not have an eigenvalue.

**3.11.** Let **A** be a symmetric, positive definite tensor on a two-dimensional real Euclidean space. For any **x** such that $|\mathbf{x}| = 1$, let $\mathbf{y} = \mathbf{Ax}$. Let $\mathbf{e}_1$, $\mathbf{e}_2$ be eigenvectors corresponding to the positive eigenvalues $\lambda_1$, $\lambda_2$ of **A**. Resolve **x** and **y** into components in the orthonormal basis **e**: $\mathbf{x} = \xi_j\,\mathbf{e}_j$, $\mathbf{y} = \eta_j\,\mathbf{e}_j$. Show that $(\eta_1/\lambda_1)^2 + (\eta_2/\lambda_2)^2 = 1$, asserting that image points $\mathbf{y} = \mathbf{Ax}$ of points **x** on the unit circle lie on an ellipse whose principal axes are directed along the eigenvectors $\mathbf{e}_1$, $\mathbf{e}_2$, and whose corresponding semi-axes have respective lengths $\lambda_1$, $\lambda_2$.

**3.12.** Let R be a real Euclidean space whose dimension need not be finite. Let $\mathbf{e}_1, \mathbf{e}_2, ..., \mathbf{e}_k$ be k orthonormal vectors in R, and put M = span $(\mathbf{e}_1, ..., \mathbf{e}_k)$. Define a linear transformation **P** on R by $\mathbf{Px} = \sum_{j=0}^{k} (\mathbf{x}, \mathbf{e}_j)\,\mathbf{e}_j$ for every **x** in R. **P** is called the *projection* of R on M. (a) Show that M is an invariant linear manifold for **P**. (b) Show that $(\mathbf{Px}, \mathbf{x} - \mathbf{Px}) = 0$ for every **x** in R. (c) Let **n** be a unit vector. Show that $\mathbf{n}\otimes\mathbf{n}$ is a projection. What is M in this case? (d) Establish *Bessel's inequality*: for any **x** in R, $|\mathbf{x}|^2 \geq |\mathbf{Px}|^2 = \sum_{j=1}^{k}(\mathbf{x}, \mathbf{e}_j)^2$. (e) Suppose that R is infinite dimensional, and let $\mathbf{e}_1, \mathbf{e}_2, ...$ be an *infinite* sequence of unit vectors in which every pair of distinct vectors is orthogonal. For any **x** in R, show that the series $\sum_{j=1}^{\infty}(\mathbf{x}, \mathbf{e}_j)^2$ converges.

**3.13.** *The Fredholm alternative for symmetric tensors.* Let **A** be a symmetric tensor on a real Euclidean space R of finite dimension n, and assume that the eigenvalues $\lambda_1, \lambda_2, ..., \lambda_n$ and corresponding orthonormal eigenvectors $\mathbf{e}_1, \mathbf{e}_2, ..., \mathbf{e}_n$ of **A** are known. Let **b** be a given vector in R, and consider the problem of finding all vectors **x** in R, if any, such that $\mathbf{Ax} = \mathbf{b}$. Let $b_i = (\mathbf{b}, \mathbf{e}_i)$, and seek solutions in the form $\mathbf{x} = \sum_{i=1}^{n} \xi_i \mathbf{e}_i$. (a) Show that the problem reduces to finding $\xi_i$s such that $\lambda_i \xi_i = b_i$, (no sum on i), i = 1, ..., n. (b) Suppose first that $\lambda_i \neq 0$ for i = 1, ..., n, i.e., **A** is non-singular. Show that the original problem has the unique solution $\mathbf{x} = \sum_{i=1}^{n} b_i \lambda_i^{-1} \mathbf{e}_i$. (c) Suppose next that the first $k \leq n$ eigenvalues of **A** vanish: $\lambda_i = 0$, i = 1, ..., k; i.e., $\lambda = 0$ is a k-fold repeated zero of the characteristic polynomial of **A**. Show that the original problem has *no* solution unless $(\mathbf{b}, \mathbf{e}_i) = 0$ for i = 1, ..., k. If these k conditions *are* satisfied, show that the solution of $\mathbf{Ax} = \mathbf{b}$ is determined only to within an arbitrary linear combination of $\mathbf{e}_1, ..., \mathbf{e}_k$. (d) Combine the results established to validate the assertion that there is a vector **x** such that $\mathbf{Ax} = \mathbf{b}$ if and only if **b** belongs to the orthogonal complement $N^{\perp}(\mathbf{A})$ of the null space $N(\mathbf{A})$ of **A**. (e) Show that the results obtained above establish the so-called *Fredholm alternative* for symmetric tensors: *either* $\mathbf{Ax} = \mathbf{b}$ has a unique solution for every **b**, in particular the unique solution $\mathbf{x} = \mathbf{o}$ for $\mathbf{b} = \mathbf{o}$, *or* the homogeneous equation $\mathbf{Ax} = \mathbf{o}$ has k linearly independent solutions, corresponding to the dimension k of $N(\mathbf{A})$. In the latter case, $\mathbf{Ax} = \mathbf{b}$ has a solution if and only if $\mathbf{b} \in N^{\perp}(\mathbf{A})$, and if this condition is satisfied, the solution of $\mathbf{Ax} = \mathbf{b}$ is determined only to within an arbitrary linear combination of k linearly independent solutions of $\mathbf{Ax} = \mathbf{o}$.

**3.14.** Let **A** be a symmetric tensor on an n-dimensional real Euclidean space $\mathsf{R}$, and let $\lambda_1 \leq \lambda_2 \leq ... \leq \lambda_n$ be the eigenvalues of **A**, arranged in algebraic order; let $\mathbf{e}_1, ..., \mathbf{e}_n$ be corresponding eigenvectors. Consider the quadratic form $f(\mathbf{x}) = (\mathbf{Ax}, \mathbf{x})$ associated with **A**. (a) If $\mathbf{x} = \xi_i \mathbf{e}_i$ is any vector in $\mathsf{R}$, show that $f(\mathbf{x}) = \sum_{i=1}^{n} \lambda_i \xi_i^2$. (b) Show that $f(\mathbf{e}_i) = \lambda_i$ for each i. (c) Suppose that **x** is a unit vector that is orthogonal to the first k − 1 of the **e**s. Show that $f(\mathbf{x}) \geq \lambda_k$. It follows that the k[th] eigenvalue of **A** can be characterized by a *minimum principle*: $\lambda_k = \min f(\mathbf{x})$, where the minimum is taken over all unit vectors **x** orthogonal to the first k − 1 eigenvectors $\mathbf{e}_1, \mathbf{e}_2, ..., \mathbf{e}_{k-1}$ of **A**. In particular, the smallest eigenvalue $\lambda_1$ (which might be negative) coincides with the minimum of $f(\mathbf{x})$ over *all* unit vectors **x**. This characterization of the eigenvalues of **A** is called *Rayleigh's principle*. For further discussion of related ideas, see the book by R. Courant and D. Hilbert [3.1] among the references listed above.

**3.15.** In the real Euclidean space $\mathsf{R}_4$ of columns of four real numbers, let **A** be the tensor whose matrix of components in the natural basis is given by

$$\underline{A} = \begin{pmatrix} 1 & -1 & 0 & 0 \\ -1 & 2 & -1 & 0 \\ 0 & -1 & 2 & -1 \\ 0 & 0 & -1 & 1 \end{pmatrix}.$$

Use Rayleigh's principle to find an upper bound for the smallest eigenvalue of **A**. How would you try to assess the quality of this bound without finding the eigenvalue exactly?

**3.16.** A linear transformation of a real Euclidean space R into itself is called a *contraction* if it shortens every vector in R in the sense that there is a real number $\alpha < 1$ such that $|\mathbf{Ax}| \leq \alpha\, |\mathbf{x}|$ for every $\mathbf{x}$ in R. Suppose that R is finite dimensional. Show that **A** is a contraction if and only if the largest eigenvalue of the symmetric tensor $\mathbf{A}^T\mathbf{A}$ is less than one.

**3.17.** Let **A** and **B** be two symmetric tensors whose matrices of components in a given orthonormal basis are given by

$$\mathrm{A} = \begin{pmatrix} 1 & 4 \\ 4 & 3 \end{pmatrix}, \quad \mathrm{B} = \begin{pmatrix} 0 & 2 \\ 2 & 1 \end{pmatrix}.$$

Do **A** and **B** have a common principal basis?

**3.18.** Show that if a Cartesian tensor **Q** satisfies (3.25), it is orthogonal.

**3.19.** Let M be a two-dimensional invariant linear manifold for the orthogonal tensor **Q** corresponding to a complex zero $\lambda = \exp(i\varphi)$ of the characteristic polynomial of **Q**. Show that **M** is also an invariant linear manifold for $\mathbf{Q}^T$ corresponding to the zero $\lambda = \exp(-i\varphi)$ of the characteristic polynomial of $\mathbf{Q}^T$.

**3.20.** Let $q_{ij}$ be the components of an orthogonal tensor in an arbitrary orthonormal basis. Show that $|q_{ij}| \leq 1$.

**3.21.** Show that the algebraic and geometric multiplicities of an eigenvalue of an orthogonal tensor coincide.

**3.22.** Let **V** be a symmetric tensor; if **R** is *any* tensor, show that $\mathbf{U} = \mathbf{R}^T\mathbf{V}\mathbf{R}$ is also symmetric. Show that the right stretch tensor **U** and the left stretch tensor **V** in the polar decomposition have the same eigenvalues.

**3.23.** Let **A** be a tensor on the real Euclidean space $\mathrm{R}_2$, and suppose that its matrix $\underline{\mathrm{A}}$ in the natural basis is given by

$$\underline{\mathrm{A}} = \begin{pmatrix} 1 & 2 \\ -2 & 2 \end{pmatrix}.$$

Find both polar decompositions of **A**.

**3.24.** A tensor **A** is orthogonal, symmetric, and positive definite. What is it?

**3.25.** Show that $\mathrm{Tr}\, \mathbf{A}^T = \mathrm{Tr}\, \mathbf{A}$.

**3.26.** Let **A** be a skew-symmetric tensor on a finite dimensional real Euclidean space. If **1** is the identity tensor, show that $\mathbf{1} + \mathbf{A}$ is nonsingular and that $(\mathbf{1} - \mathbf{A})(\mathbf{1} + \mathbf{A})^{-1}$ is orthogonal.

**3.27.** *Square root of a symmetric, non-negative definite tensor.* Let **A** be a symmetric, non-negative definite tensor on an n-dimensional real Euclidean space R, and suppose that its first k eigenvalues are zero, with the remaining ones positive: $\lambda_1 = \cdots = \lambda_k = 0, 0 < \lambda_{k+1} \leq \lambda_{k+2} \cdots \leq \lambda_n$. Show that **A** has a unique non-negative definite symmetric square root **B** by modifying the argument used in the foregoing chapter to establish the uniqueness and existence of the positive definite square root of a positive definite tensor. Suppose first that there *is* such a **B**. (a) If $\mathbf{e}_{k+1}, \ldots, \mathbf{e}_n$ are the eigenvectors of **A** corresponding to its positive eigenvalues, the argument given in the text can be used to show that these vectors are again eigenvectors of **B** corresponding to eigenvalues $\lambda_{k+1}^{1/2}, \ldots, \lambda_n^{1/2}$. (b) Let $\mathsf{M} = \text{span}\ (\mathbf{e}_{k+1}, \ldots, \mathbf{e}_n)$, and let $\mathsf{M}^{\perp}$ be its orthogonal complement. If $\mathbf{f} \in \mathsf{M}^{\perp}$, show that $(\mathbf{Bf}, \mathbf{e}_i) = 0$ for $i = k + 1, \ldots, n$, so that **Bf** is also in $\mathsf{M}^{\perp}$. (c) Let $\mathbf{e}_1, \ldots, \mathbf{e}_k$ be the k orthonormal eigenvectors of **A** corresponding to the k-fold repeated eigenvalue $\lambda = 0$: $\mathbf{Ae}_i = \mathbf{o}$, $i = 1, \ldots, k$. Show that $\text{span}\ (\mathbf{e}_1, \ldots, \mathbf{e}_k) = \mathsf{M}^{\perp}$, and that $\mathbf{Af} = \mathbf{o}$ for every vector **f** in $\mathsf{M}^{\perp}$. (d) Since the restriction of **B** to $\mathsf{M}^{\perp}$ is a symmetric linear transformation of $\mathsf{M}^{\perp}$ into itself, there is an orthonormal basis $\mathbf{f}_1, \ldots, \mathbf{f}_k$ for $\mathsf{M}^{\perp}$ for which $\mathbf{Bf}_i = \mu_i \mathbf{f}_i$ (no sum on i), $i = 1, \ldots, k$. (e) The vectors $\mathbf{f}_1, \ldots \mathbf{f}_k, \mathbf{e}_{k+1}, \ldots, \mathbf{e}_n$ comprise an orthonormal basis **g** for R, and in this basis, the matrix B of components of **B** is diagonal with entries $\mu_1, \ldots, \mu_k, \lambda_{k+1}^{1/2}, \ldots, \lambda_n^{1/2}$ along the principal diagonal. The matrix A of **A** in this basis is also diagonal, with diagonal entries $0, \ldots, 0, \lambda_{k+1}, \ldots, \lambda_n$. (f) Since $B^2 = A$ by assumption, necessarily $\mu_1 = \cdots = \mu_k = 0$. Thus if a tensor **B** with the assumed properties exists, then it is necessarily the tensor whose matrix in the basis **g** is diagonal with diagonal elements $0, \cdots, 0, \lambda_{k+1}^{1/2}, \cdots \lambda_n^{1/2}$. (g) Clearly the tensor **B** specified in this way is symmetric, non-negative definite and has the property $\mathbf{B}^2 = \mathbf{A}$, establishing the existence and uniqueness of the non-negative definite symmetric square root **B** of a non-negative definite symmetric tensor **A**.

**3.28.** *Generalization of the polar decomposition.* Let **A** be *any* tensor on a finite dimensional real Euclidean space. Then there are unique symmetric tensors **U** and **V**, each non-negative definite, and orthogonal tensors **Q** and **R** such that $\mathbf{A} = \mathbf{QU} = \mathbf{VR}$. Unless **A** is non-singular, neither **Q** nor **R** need be unique, and thus it need not be true that $\mathbf{Q} = \mathbf{R}$, as is the case when **A** is non-singular. To establish this theorem, prove the results asserted in the following steps. It is sufficient to consider only the case where **A** is singular. (a) The symmetric tensor $\mathbf{A}^T\mathbf{A}$ is non-negative definite. (b) If the right polar decomposition $\mathbf{A} = \mathbf{UQ}$ exists, then necessarily **U** is uniquely determined as the non-negative definite symmetric square root of $\mathbf{A}^T\mathbf{A}$. (See the preceding problem). Let

**U** be so defined. (c) Let $\lambda_1 = \cdots = \lambda_k = 0$ be the k-fold repeated zero eigenvalue of **U**, and let $\lambda_{k+1}, \cdots, \lambda_n$ be its positive eigenvalues. Let $\mathbf{e}_1, ..., \mathbf{e}_k, \mathbf{e}_{k+1}, ..., \mathbf{e}_n$ be n orthonormal eigenvectors of **U** corresponding to $\lambda_1, ..., \lambda_k, \lambda_{k+1}, ..., \lambda_n$, respectively, and define n vectors $\mathbf{f}_i$ by $\mathbf{f}_i = \mathbf{A}\mathbf{e}_i$, i = 1, ..., n. Show that $(\mathbf{f}_i, \mathbf{f}_j) = \lambda_i\lambda_j\, \delta_{ij}$, no sum on i or j. Thus in particular, $\mathbf{f}_i = \mathbf{o}$ for i = 1, ..., k, so that

$$\mathbf{A}\mathbf{e}_i = \mathbf{o},\ i = 1, ..., k. \quad (*)$$

(d) For i = k + 1, ..., n, let $\mathbf{g}_i = \lambda_i^{-1}\mathbf{f}_i$ (no sum), and show that the **g**s are orthonormal. Let $\mathsf{M} = \text{span}\,(\mathbf{g}_{k+1}, ..., \mathbf{g}_n)$, and let $\mathsf{M}^{\perp}$ be its orthogonal complement. Show that the k-dimensional space $\mathsf{M}^{\perp}$ is the null space of **U**. (e) Let $\mathbf{h}_1, ..., \mathbf{h}_k$ be an *arbitrary* orthonormal basis for $\mathsf{M}^{\perp}$, so that $\mathbf{h}_1, ..., \mathbf{h}_k, \mathbf{e}_{k+1}, ..., \mathbf{e}_n$ forms an orthonormal basis for $\mathsf{R}$. Define a tensor **Q** by specifying its values on this basis as follows:

$$\mathbf{Q}\mathbf{h}_i = \mathbf{h}_i,\ i = 1, ..., k, \quad \mathbf{Q}\mathbf{e}_i = \mathbf{g}_i,\ i = k + 1, ..., n;$$

show that $|\mathbf{Q}\mathbf{x}| = |\mathbf{x}|$ for every **x** in $\mathsf{R}$, so **Q** is indeed orthogonal. (f) Finally, by representing **x** in the basis $\mathbf{h}_1, ..., \mathbf{h}_k, \mathbf{e}_{k+1}, ..., \mathbf{e}_n$ and making use of (*), prove that $\mathbf{QUx} = \mathbf{Ax}$ for every **x** in $\mathsf{R}$, thus establishing the right polar decomposition $\mathbf{A} = \mathbf{QU}$. Note that in the special case where **A** is the null tensor **O**, one has $\mathbf{U} = \mathbf{O}$ as well, and $\mathbf{A} = \mathbf{QU}$ holds for *any* orthogonal tensor **Q**, showing that **Q** need not be unique when **A** is singular. The left polar decomposition $\mathbf{A} = \mathbf{VR}$ is established by an analogous argument.

**3.29.** *The Fredholm alternative for an arbitrary tensor.* Let **R** be an n-dimensional real Euclidean space, and let **A** be a given tensor, **b** a given vector. Consider the problem of finding all vectors **x** in $\mathsf{R}$ such that $\mathbf{Ax} = \mathbf{b}$. (a) Suppose this problem has a solution **x**. Let **y** be in the null space $N(\mathbf{A}^T)$ of $\mathbf{A}^T$; show that $(\mathbf{b}, \mathbf{y}) = 0$, so that **b** is in $N^{\perp}(\mathbf{A}^T)$. (b) Conversely, assume that $\mathbf{b} \in N^{\perp}(\mathbf{A})$. By the version of the polar decomposition established in the preceding problem, one can find an orthogonal tensor **R** and a symmetric, non-negative definite tensor **V** such that $\mathbf{A} = \mathbf{VR}$. Let **z** be in the null space $N(\mathbf{V})$ of **V**. Show that $\mathbf{A}^T\mathbf{z} = \mathbf{o}$, so **z** is in the null space of $\mathbf{A}^T$ as well. Hence $(\mathbf{b}, \mathbf{z}) = 0$ for every vector **z** in $N(\mathbf{V})$. Use the results established in connection with the Fredholm alternative for *symmetric* tensors (Problem 3.13) to prove that there is a vector **w** in $\mathsf{R}$ such that $\mathbf{Vw} = \mathbf{b}$. Next, set $\mathbf{x} = \mathbf{R}^T\mathbf{w}$ and show that $\mathbf{Ax} = \mathbf{b}$, so that the problem $\mathbf{Ax} = \mathbf{b}$ has a solution. Thus a necessary and sufficient condition for the existence of a solution **x** of $\mathbf{Ax} = \mathbf{b}$ is that **b** belong to the orthogonal complement of the null space of $\mathbf{A}^T$. This establishes the *Fredholm alternative*: *either* $\mathbf{Ax} = \mathbf{b}$ has a unique solution for every $\mathbf{b}$, in particular the unique solution $\mathbf{x} = \mathbf{o}$ when $\mathbf{b} = \mathbf{o}$, *or* the homogeneous equation $\mathbf{Ax} = \mathbf{o}$ has k linearly independent solutions, corresponding to the dimension k of $N(\mathbf{A})$. In the latter case, $N(\mathbf{A}^T)$ has dimension k as well, and $\mathbf{Ax} = \mathbf{b}$ has a solution if and only if $\mathbf{b} \in N^{\perp}(\mathbf{A}^T)$. If this condition is satisfied, the solution of $\mathbf{Ax} = \mathbf{b}$ is determined only to within an arbitrary linear combination of k linearly independent solutions of $\mathbf{Ax} = \mathbf{o}$.

**3.30.** Let $f(\mathbf{x}) = (\mathbf{x}, \mathbf{Ax})$ be the quadratic form associated with the positive definite symmetric tensor $\mathbf{A}$. By representing $f(\mathbf{x})$ in in arbitrary orthonormal basis $\mathbf{e}$ and choosing $\mathbf{x}$ appropriately, show that the diagonal components $a^{\mathbf{e}}_{ii}$ (no sum on i) of $\mathbf{A}$ must be positive for any such basis $\mathbf{e}$.

CHAPTER 4

# 4-TENSORS

We now make use of the ideas introduced in the previous chapters to explore the concept of *4-tensors*, which are of great importance in the continuum mechanics of solids. The role of 4-tensors in this subject is well illustrated in the article by M.E. Gurtin [4.1] listed among the references at the end of this chapter.

Let R be an n-dimensional real Euclidean space, and let L be the collection of all linear transformations of R into itself, i.e., L is the set of all tensors. By adducing the natural notions of addition of two tensors and multiplication of a tensor by a scalar, we may render L a real linear vector space in its own right. Suppose in addition there is a scalar product on L, conforming to the fundamental rules for such products laid down in Chapter 1; let us denote the scalar product of two tensors **A** and **B** in L by the special symbol $\langle \mathbf{A}, \mathbf{B} \rangle$. Thus L has now become a real Euclidean space. In a moment, we shall show that there is a particular scalar product in L that is *natural* in a definite sense.

Let $\mathbf{e}_1, \mathbf{e}_2, ..., \mathbf{e}_n$ be an orthonormal basis for the underlying space R, and define $n^2$ tensors $\mathbf{E}_{ij}$ in L through the tensor products of pairs of vectors of the basis **e**:

$$\mathbf{E}_{ij} = \mathbf{e}_i \otimes \mathbf{e}_j. \tag{4.1}$$

A moment's reflection shows that the matrix $\underline{E}^{\mathbf{e}}_{ij}$ (watch out for bumpy notation as we go along!) of components of $\mathbf{E}_{ij}$ in the basis **e** has a *one* in the $i^{th}$ row, $j^{th}$ column, and *zeros* everywhere else. Now suppose that **A** is *any* tensor in L, and let $\underline{A}^{\mathbf{e}} = (a^{\mathbf{e}}_{ij})$ be its matrix of components in **e**. Clearly, $\underline{A}^{\mathbf{e}} = a^{\mathbf{e}}_{ij}\, \underline{E}^{\mathbf{e}}_{ij}$; remember the summation convention. From this it follows immediately that any tensor **A** in L may be represented in the form

$$\mathbf{A} = a^{\mathbf{e}}_{ij}\, \mathbf{E}_{ij}. \tag{4.2}$$

If we knew that the $n^2$ tensors $\mathbf{E}_{ij}$ were linearly independent, we could immediately infer from (4.2) that the $\mathbf{E}_{ij}$s comprise a basis for L, whose dimension would thus be $n^2$. To confirm this linear independence, suppose that there are $n^2$ constants $\alpha_{ij}$ such that

$$\alpha_{ij}\, \mathbf{E}_{ij} = \alpha_{ij}\, \mathbf{e}_i \otimes \mathbf{e}_j = \mathbf{O}. \tag{4.3}$$

By feeding the vector $\mathbf{e}_k$ to the tensors on either side of $(4.3)_2$, one finds that

$$\alpha_{ik}\, \mathbf{e}_i = \mathbf{o}, \tag{4.4}$$

for each k = 1, ..., n. But the linear independence of the $\mathbf{e}_i$s then implies that $\alpha_{ik} = 0$, for all values of i and k. Thus the $\mathbf{E}_{ij}$s are indeed linearly independent; we say they form the basis in L that is *induced* from the underlying basis **e** in R. Of course, there are other bases for L as well.

Suppose it so happens that the induced basis **E** is orthonormal with respect to the given scalar product $\langle \cdot, \cdot \rangle$: thus $\langle \mathbf{E}_{ij}, \mathbf{E}_{kl} \rangle$ takes the value zero when $i \neq k$ and $j \neq l$, and the value one when $i = k$ and $j = l$. With the help of the Kronecker delta, this orthonormality may be expressed as

$$\langle \mathbf{E}_{ij}, \mathbf{E}_{kl} \rangle = \delta_{ik}\, \delta_{jl}. \tag{4.5}$$

Now pick two tensors **A** and **B** in L and represent them in the form (4.2) for the purpose of computing their scalar product. Bearing the summation convention in mind and using (4.5), one finds that

$$\langle \mathbf{A}, \mathbf{B} \rangle = a_{ij}\, b_{kl}\, \langle \mathbf{E}_{ij}, \mathbf{E}_{kl} \rangle = a_{ij} b_{kl} \delta_{ik} \delta_{jl} = a_{ij}\, b_{ij} = \mathrm{Tr}\, (\mathbf{A}\mathbf{B}^T). \tag{4.6}$$

Thus if the induced basis **E** is orthonormal, the scalar product $\langle \mathbf{A}, \mathbf{B} \rangle$ *must be computed according to the recipe derived in (4.6)*: $\langle \mathbf{A}, \mathbf{B} \rangle = \mathrm{Tr}\, (\mathbf{A}\mathbf{B}^T)$. It is easy to show that this prescription of a scalar product conforms to the rules governing such products as set out in Chapter 1.

Conversely, suppose the scalar product is defined as indicated by the extreme members of (4.6). It is then an easy exercise to show that (4.5) holds, so that the induced basis **E** is indeed orthonormal; see Problem 4.1. Thus we have shown that the induced basis is orthonormal *if and only if* the scalar product of two tensors **A** and **B** is that defined by

$$\langle \mathbf{A}, \mathbf{B} \rangle = \mathrm{Tr}\, (\mathbf{A}\mathbf{B}^T). \tag{4.7}$$

The notion of the "length" $\|\mathbf{A}\|$ of a tensor **A** in L inherited from this scalar product is that defined by

$$\|\mathbf{A}\| = (\mathrm{Tr}\, (\mathbf{A}\mathbf{A}^T))^{1/2} = (a_{ij}\, a_{ij})^{1/2}, \tag{4.8}$$

where the $a_{ij}$s are the components of **A** in any orthonormal basis. Thus the particular scalar product (4.7) is natural for L in the foregoing sense; we assume it to be in force from here on.

Since L is a vector space, we may consider linear transformations of L into itself. If $\underline{\underline{\mathbf{C}}}$ is such a transformation, we call it a *4-tensor*, for reasons that will show up a little later. As a linear transformation, a 4-tensor accepts as input any (ordinary) tensor **A** and delivers another (ordinary) tensor **B** as output:

$$\mathbf{B} = \underline{\underline{\mathbf{C}}}\ \mathbf{A}. \tag{4.9}$$

We shall consistently use a doubly-underlined boldface letter to designate a 4-tensor. To emphasize the distinction between 4-tensors and ordinary tensors, when necessary for clarity we shall refer to the latter as *2-tensors*; this additional piece of mysterious terminology will be explained later, too.

In the mechanics of elastic solids, a relation of the form (4.9) figures prominently; it relates the *strain* **A** to the *stress* **B**; $\underline{\underline{\mathbf{C}}}$ is called the *elasticity tensor*.

***Example 4.1.*** *Some special 4-tensors.* There is of course an *identity* 4-tensor $\underline{\underline{\mathbf{1}}}$ for which $\underline{\underline{\mathbf{1}}}\ \mathbf{A} = \mathbf{A}$ for every 2-tensor **A** in L. In addition, one has the *transposition* 4-tensor $\underline{\underline{\mathbf{T}}}$ defined by $\underline{\underline{\mathbf{T}}}\ \mathbf{A} = \mathbf{A}^{\mathrm{T}}$, the *symmetrizer* $\underline{\underline{\mathbf{S}}}$ defined by $\underline{\underline{\mathbf{S}}}\mathbf{A} = (1/2)(\mathbf{A} + \mathbf{A}^{\mathrm{T}}) = (1/2)(\underline{\underline{\mathbf{1}}} + \underline{\underline{\mathbf{T}}})\mathbf{A}$ and the *skew-symmetrizer* $\underline{\underline{\mathbf{W}}}$ defined by $\underline{\underline{\mathbf{W}}}\ \mathbf{A} = (1/2)(\mathbf{A} - \mathbf{A}^{\mathrm{T}}) = (1/2)(\underline{\underline{\mathbf{1}}} - \underline{\underline{\mathbf{T}}})\mathbf{A}$. It is easily seen that the operations upon 2-tensors performed by $\underline{\underline{\mathbf{1}}}$, $\underline{\underline{\mathbf{T}}}$, $\underline{\underline{\mathbf{S}}}$, and $\underline{\underline{\mathbf{W}}}$ are indeed linear.

***Example 4.2.*** *The 4-tensor product of two 2-tensors.* Let **B** and **C** be two fixed 2-tensors. By analogy with the notion of the tensor product of two vectors, define a 4-tensor $\underline{\underline{\mathbf{D}}}$ by $\underline{\underline{\mathbf{D}}}\mathbf{A} = \langle \mathbf{C}, \mathbf{A}\rangle\, \mathbf{B}$ for every 2-tensor **A**. $\underline{\underline{\mathbf{D}}}$ is called the 4-tensor product of the 2-tensors **B** and **C**, and one writes $\underline{\underline{\mathbf{D}}} = \mathbf{B}\otimes\mathbf{C}$.

The notion of components of a 4-tensor is introduced in the same way that components of 2-tensors were defined in (2.25). Technically, perhaps we should speak of the components of $\underline{\underline{\mathbf{C}}}$ in the basis **E**, but it is traditional instead to speak of the components $c^{\mathbf{e}}_{ijkl}$ of $\underline{\underline{\mathbf{C}}}$ in the basis **e** of the underlying vector space R that gives rise to the induced basis **E**. These are defined through

$$\underline{\underline{\mathbf{C}}}\ \mathbf{E}_{kl} = c^{\mathbf{e}}_{ijkl}\, \mathbf{E}_{ij}; \tag{4.10}$$

the summation convention is in force. As usual, we omit the superscript **e**, writing instead $c_{ijkl}$, unless confusion would result. A 4-tensor has $n^4$ scalar components. In view of the orthonormality of the $\mathbf{E}_{ij}$s with respect to the scalar product in L, it follows from (4.10) that

$$c_{ijkl} = \langle \mathbf{E}_{ij}, \underline{\underline{\mathbf{C}}}\,\mathbf{E}_{kl}\rangle; \tag{4.11}$$

note the nice analogy with (3.1).

In Chapter 2, we showed that if the vectors **x** and **y** are related by **y** = **Ax**, where **A** is a 2-tensor, then the components of **x** and **y** are related through the components of **A** by (2.28). Suppose now that the 2-tensors **A** and **B** are related by (4.9); what is the corresponding statement in terms of components? Representing all the players in (4.9) in terms of components yields

$$b_{ij}\,\mathbf{E}_{ij} = \underline{\underline{\mathbf{C}}}\,(a_{kl}\,\mathbf{E}_{kl}) = a_{kl}\,\underline{\underline{\mathbf{C}}}\,\mathbf{E}_{kl} = a_{kl}\,c_{ijkl}\mathbf{E}_{ij}; \tag{4.12}$$

by the linear independence of the $\mathbf{E}_{ij}$s, it then follows that

$$b_{ij} = c_{ijkl}\,a_{kl}, \tag{4.13}$$

which is the counterpart for 4-tensors of (2.28).

It is easy to show that, if $\underline{\underline{\mathbf{C}}}$ and $\underline{\underline{\mathbf{D}}}$ are two 4-tensors, then the components $g_{ijkl}$ of the 4-tensor $\underline{\underline{\mathbf{G}}} = \underline{\underline{\mathbf{C}}}\,\underline{\underline{\mathbf{D}}}$ are given by $g_{ijkl} = c_{ijpq}\,d_{pqkl}$.

> ***Example 4.3.*** *Components of* $\underline{\underline{\mathbf{1}}}$, $\underline{\underline{\mathbf{T}}}$, $\underline{\underline{\mathbf{S}}}$, *and* $\underline{\underline{\mathbf{W}}}$. Let us apply (4.11) when $\underline{\underline{\mathbf{C}}} = \underline{\underline{\mathbf{1}}}$ is the identity 4-tensor. From (4.11) and (4.5), the components $i_{ijkl}$ of $\underline{\underline{\mathbf{1}}}$ are given by $i_{ijkl} = \langle \mathbf{E}_{ij}, \mathbf{E}_{kl}\rangle = \delta_{ik}\,\delta_{jl}$. If $\underline{\underline{\mathbf{C}}} = \underline{\underline{\mathbf{T}}}$ is the transposition tensor, then from (4.11), (4.1), (3.9), and (4.5), $t_{ijkl} = \langle \mathbf{E}_{ij}, \underline{\underline{\mathbf{T}}}\mathbf{E}_{kl}\rangle = \langle \mathbf{E}_{ij}, \mathbf{E}_{lk}\rangle = \delta_{il}\,\delta_{jk}$ are the components of $\underline{\underline{\mathbf{T}}}$. The components of $\underline{\underline{\mathbf{S}}}$ and $\underline{\underline{\mathbf{W}}}$ are readily found to be $s_{ijkl} = (1/2)(\delta_{ik}\,\delta_{jl} + \delta_{il}\,\delta_{jk})$ and $w_{ijkl} = (1/2)(\delta_{ik}\,\delta_{jl} - \delta_{il}\,\delta_{jk})$, respectively.

Since L is a finite-dimensional Euclidean space, every 4-tensor $\underline{\underline{\mathbf{C}}}$ has a transpose $\underline{\underline{\mathbf{C}}}^{T}$ characterized by $\langle \underline{\underline{\mathbf{C}}}\mathbf{A}, \mathbf{B}\rangle = \langle \mathbf{A}, \underline{\underline{\mathbf{C}}}^{T}\mathbf{B}\rangle$ for every pair of 2-tensors **A**, **B**. A 4-tensor $\underline{\underline{\mathbf{C}}}$ is then symmetric if $\underline{\underline{\mathbf{C}}} = \underline{\underline{\mathbf{C}}}^{T}$. The transformation tensor $\underline{\underline{\mathbf{T}}}$, for example, is symmetric, because $\langle \underline{\underline{\mathbf{T}}}\mathbf{A}, \mathbf{B}\rangle = \langle \mathbf{A}^{T}, \mathbf{B}\rangle = \mathrm{Tr}\,(\mathbf{A}^{T}\mathbf{B}^{T}) = \mathrm{Tr}\,(\mathbf{B}^{T}\mathbf{A}^{T}) = \mathrm{Tr}\,((\mathbf{AB})^{T}) = \mathrm{Tr}\,(\mathbf{AB}) = \mathrm{Tr}\,(\mathbf{A}(\mathbf{B}^{T})^{T})$, whence $\langle \underline{\underline{\mathbf{T}}}\mathbf{A}, \mathbf{B}\rangle = \langle \mathbf{A}, \mathbf{B}^{T}\rangle = \langle \mathbf{A}, \underline{\underline{\mathbf{T}}}\mathbf{B}\rangle$.

Observe from (4.11) and the definition of the transpose of a 4-tensor that $c_{klij} = \langle \mathbf{E}_{kl}, \underline{\underline{\mathbf{C}}}\mathbf{E}_{ij}\rangle = \langle \mathbf{E}_{ij}, \underline{\underline{\mathbf{C}}}^{T}\mathbf{E}_{kl}\rangle$; this shows that if $c_{ijkl}$ are the components of $\underline{\underline{\mathbf{C}}}$, then $c_{klij}$ are the components of $\underline{\underline{\mathbf{C}}}^{T}$. In particular, if $\underline{\underline{\mathbf{C}}}$ is a symmetric 4-tensor, then its components satisfy $c_{ijkl} = c_{klij}$, the subscripts ij and kl being interchangeable as pairs.

From the general theory described in Chapter 3, it follows immediately that any *symmetric* 4-tensor $\underline{\underline{\mathbf{C}}}$ possesses $m = n^2$ eigenvalues $\alpha_1, \ldots, \alpha_m$ and m corresponding orthonormal 2-tensors $\mathbf{A}_1, \ldots, \mathbf{A}_m$ (the "eigentensors" of $\underline{\underline{\mathbf{C}}}$) such that

$$\underline{\underline{\mathbf{C}}}\,\mathbf{A}_i = \alpha_i\mathbf{A}_i \text{ (no sum on i) } i = 1, \ldots, m. \tag{4.14}$$

Moreover, the symmetric 4-tensor $\underline{\underline{\mathbf{C}}}$ can be represented by a spectral formula:

$$\underline{\underline{\mathbf{C}}} = \sum_{i=1}^{m} \alpha_i \, \mathbf{A}_i \otimes \mathbf{A}_i. \tag{4.15}$$

There are two other symmetry-related notions for 4-tensors. If a 4-tensor $\underline{\underline{\mathbf{C}}}$ satisfies $\underline{\underline{\mathbf{T}}}\,\underline{\underline{\mathbf{C}}} = \underline{\underline{\mathbf{C}}}$, where $\underline{\underline{\mathbf{T}}}$ is the transposition 4-tensor, then $\underline{\underline{\mathbf{C}}}$ is said to have the *first minor symmetry*. If the 4-tensor $\underline{\underline{\mathbf{C}}}$ in the relation (4.9) has the first minor symmetry, then (4.9) implies that $\mathbf{B}^T = \underline{\underline{\mathbf{T}}}\mathbf{B} = \underline{\underline{\mathbf{T}}}\,\underline{\underline{\mathbf{C}}}\,\mathbf{A} = \underline{\underline{\mathbf{C}}}\,\mathbf{A} = \mathbf{B}$, so that the output $\mathbf{B}$ of the 4-tensor $\underline{\underline{\mathbf{C}}}$ in (4.9) is always symmetric, regardless of the nature of the input $\mathbf{A}$. It is easy to see that, if $\underline{\underline{\mathbf{C}}}$ has the first minor symmetry, then its components satisfy $c_{ijkl} = c_{jikl}$. On the other hand, if the 4-tensor $\underline{\underline{\mathbf{C}}}$ satisfies $\underline{\underline{\mathbf{C}}}\,\underline{\underline{\mathbf{T}}} = \underline{\underline{\mathbf{C}}}$, the $\underline{\underline{\mathbf{C}}}$ is said to have the *second minor symmetry*. If $\underline{\underline{\mathbf{C}}}$ in (4.9) has this property, it follows from (4.9) that $\mathbf{B} = (1/2)\,(\underline{\underline{\mathbf{C}}}\,\mathbf{A} + \underline{\underline{\mathbf{C}}}\,\underline{\underline{\mathbf{T}}}\,\mathbf{A}) = \underline{\underline{\mathbf{C}}}((1/2)(\mathbf{A} + \mathbf{A}^T))$; thus the output 2-tensor $\mathbf{B}$ in the relation (4.9) depends only on the symmetric part $(1/2)(\mathbf{A} + \mathbf{A}^T)$ of the input 2-tensor $\mathbf{A}$. When a 4-tensor $\underline{\underline{\mathbf{C}}}$ has the second minor symmetry, its components satisfy $c_{ijkl} = c_{ijlk}$.

If a 4-tensor $\underline{\underline{\mathbf{C}}}$ is symmetric in the fundamental sense that $\underline{\underline{\mathbf{C}}} = \underline{\underline{\mathbf{C}}}^T$, one says that $\underline{\underline{\mathbf{C}}}$ has the *major symmetry* when it is important to emphasize the distinction between *this* notion of symmetry and one or both of the *minor* symmetries. In Problem 4.5 it is shown that if a 4-tensor has the major symmetry and one of the minor symmetries, it also has the other minor symmetry.

Naturally, a 4-tensor $\underline{\underline{\mathbf{Q}}}$ is orthogonal if it preserves length in $\mathsf{L}$: $\|\underline{\underline{\mathbf{Q}}}\,\mathbf{A}\| = \|\mathbf{A}\|$ for every 2-tensor $\mathbf{A}$ in $\mathsf{L}$; here the notion of length is that of (4.8). An orthogonal 4-tensor will also preserve the scalar product in $\mathsf{L}$, and such a tensor of course satisfies

$$\underline{\underline{\mathbf{Q}}}\,\underline{\underline{\mathbf{Q}}}^T = \underline{\underline{\mathbf{Q}}}^T\underline{\underline{\mathbf{Q}}} = \underline{\underline{\mathbf{1}}} \tag{4.16}$$

***Example 4.4.*** *A special class of orthogonal 4-tensors: the 4-rotations.* Let $\mathbf{Q}$ be a fixed orthogonal 2-tensor, and define a 4-tensor $\underline{\underline{\mathbf{Q}}}$ by

$$\underline{\underline{\mathbf{Q}}}\,\mathbf{A} = \mathbf{Q}\mathbf{A}\mathbf{Q}^T \text{ for every 2-tensor } \mathbf{A} \text{ in } \mathsf{L}. \tag{4.17}$$

To see that the 4-tensor $\underline{\underline{\mathbf{Q}}}$ preserves length in $\mathsf{L}$ and therefore is orthogonal, observe that $\|\underline{\underline{\mathbf{Q}}}\,\mathbf{A}\|^2 = \langle \underline{\underline{\mathbf{Q}}}\,\mathbf{A},\ \underline{\underline{\mathbf{Q}}}\,\mathbf{A}\rangle = \mathrm{Tr}\ (\mathbf{Q}\mathbf{A}\mathbf{Q}^T(\mathbf{Q}\mathbf{A}\mathbf{Q}^T)^T) = \mathrm{Tr}\ (\mathbf{Q}\mathbf{A}\mathbf{A}^T\mathbf{Q}^T)$; invoking the property (3.42) of the trace, one finds that $\|\underline{\underline{\mathbf{Q}}}\,\mathbf{A}\|^2 = \mathrm{Tr}\ (\mathbf{A}\mathbf{A}^T) = \|\mathbf{A}\|^2$, establishing the orthogonality of the 4-tensor $\underline{\underline{\mathbf{Q}}}$ defined by (4.17).

To find the components $q_{ijkl}$ in the basis $\mathbf{e}$ of an orthogonal 4-tensor $\underline{\underline{\mathbf{Q}}}$ of the special form defined by (4.17), we need to compute $\underline{\underline{\mathbf{Q}}}\,\mathbf{E}_{kl}$ as a linear combination of the $\mathbf{E}_{ij}$s. We begin by observing that, for any $\mathbf{x}$ in the underlying vector space $\mathsf{R}$, we have $(\underline{\underline{\mathbf{Q}}}\,\mathbf{E}_{kl})\,\mathbf{x} = \mathbf{Q}\mathbf{E}_{kl}\mathbf{Q}^T\mathbf{x} = \mathbf{Q}((\mathbf{e}_k \otimes \mathbf{e}_l)(\mathbf{Q}^T\mathbf{x})) =$

$\mathbf{Q}\,(\mathbf{e}_k\,(\mathbf{e}_l, \mathbf{Q}^T\mathbf{x})) = \mathbf{Q}\,(\mathbf{e}_k\,(\mathbf{Q}\mathbf{e}_l, \mathbf{x})) = (\mathbf{Q}\mathbf{e}_l, \mathbf{x})\,\mathbf{Q}\mathbf{e}_k = ((\mathbf{Q}\mathbf{e}_k)\otimes(\mathbf{Q}\mathbf{e}_l))\,\mathbf{x}$, so that $\underline{\underline{\mathbf{Q}}}\,\mathbf{E}_{kl} = (\mathbf{Q}\mathbf{e}_k)\otimes(\mathbf{Q}\mathbf{e}_l)$. In terms of the components $q_{ij}$ of the 2-tensor $\mathbf{Q}$ entering the definition (4.17), we may write this result in the form $\underline{\underline{\mathbf{Q}}}\,\mathbf{E}_{kl} = (q_{ik}\,\mathbf{e}_i)\otimes(q_{jl}\,\mathbf{e}_j) = q_{ik}\,q_{jl}\,\mathbf{e}_i\otimes\mathbf{e}_j = q_{ik}\,q_{jl}\,\mathbf{E}_{ij}$. But by definition of the components of a 4-tensor, $\underline{\underline{\mathbf{Q}}}\,\mathbf{E}_{kl} = q_{ijkl}\,\mathbf{E}_{ij}$; comparing the last two formulas shows that

$$q_{ijkl} = q_{ik}\,q_{jl}. \tag{4.18}$$

To understand the geometric significance of orthogonal 4-tensors of the special form (4.17), it is useful to introduce the idea of *rotation of a 2-tensor*. Let $\mathbf{A}$ and $\mathbf{B}$ be two 2-tensors, and let $\mathbf{e}_1, \ldots, \mathbf{e}_n$ and $\mathbf{f}_1, \ldots, \mathbf{f}_n$ be two orthonormal bases in $\mathsf{R}$ that are related by

$$\mathbf{f}_i = \mathbf{R}\mathbf{e}_i,\ i = 1, \ldots, n \tag{4.19}$$

where $\mathbf{R}$ is a *proper* orthogonal 2-tensor. Suppose it happens that the components in $\mathbf{e}$ of $\mathbf{A}$ and the components in $\mathbf{f}$ of $\mathbf{B}$ are the same: $b^{\mathbf{f}}_{ij} = a^{\mathbf{e}}_{ij}$, or in terms of the corresponding matrices of components, $\underline{B}^{\mathbf{f}} = \underline{A}^{\mathbf{e}}$. Then we say that $\mathbf{A}$ has been *rotated to* $\mathbf{B}$ *by means of* $\mathbf{R}$. By the change-of-basis formula (3.40) for the components of a 2-tensor, $\underline{B}^{\mathbf{f}} = (\underline{R}^{\mathbf{e}})^T\underline{B}^{\mathbf{e}}\,\underline{R}^{\mathbf{e}}$; thus if $\mathbf{A}$ has been rotated to $\mathbf{B}$, we have $\underline{A}^{\mathbf{e}} = (\underline{R}^{\mathbf{e}})^T\underline{B}^{\mathbf{e}}\,\underline{R}^{\mathbf{e}}$, or

$$\mathbf{B} = \mathbf{R}\mathbf{A}\mathbf{R}^T. \tag{4.20}$$

If $\underline{\underline{\mathbf{R}}}$ is the orthogonal 4-tensor defined by $\underline{\underline{\mathbf{R}}}\,\mathbf{G} = \mathbf{R}\mathbf{G}\mathbf{R}^T$ for every 2-tensor $\mathbf{G}$, then the relationship between $\mathbf{A}$ and $\mathbf{B}$ may be expressed in 4-tensor language as

$$\mathbf{B} = \underline{\underline{\mathbf{R}}}\,\mathbf{A}. \tag{4.21}$$

This explains why we call an orthogonal 4-tensor $\underline{\underline{\mathbf{Q}}}$ of the special form (4.17) a *4-rotation* if the generating orthogonal 2-tensor $\mathbf{Q}$ is proper.

It is important to observe that not *all* orthogonal 4-tensors are of the form of those introduced in Example 4.4; this is the subject of Problem 4.6.

We turn now to the change-of basis formula for the components of a 4-tensor. As in the discussion leading to the change-of-basis formula (3.40) for the components of a 2-tensor, let $\mathbf{e}_1, \mathbf{e}_2, \ldots, \mathbf{e}_n$ and $\mathbf{f}_1, \mathbf{f}_2, \ldots, \mathbf{f}_n$ be two orthonormal bases related by (4.19), where $\mathbf{R}$ is a proper orthogonal 2-tensor. Let $\mathbf{E}_{ij} = \mathbf{e}_i\otimes\mathbf{e}_j$ and $\mathbf{F}_{ij} = \mathbf{f}_i\otimes\mathbf{f}_j$ be the 2-tensors comprising the induced bases in $\mathsf{L}$ associated with the respective bases $\mathbf{e}$ and $\mathbf{f}$ in $\mathsf{R}$. Our first task is to ascertain the relation between the $\mathbf{E}_{ij}$s and the $\mathbf{F}_{ij}$s. We begin by noting that $\mathbf{F}_{ij} = (\mathbf{R}\mathbf{e}_i)\otimes(\mathbf{R}\mathbf{e}_j)$; thus for any vector $\mathbf{x}$ in $\mathsf{R}$ thought of as an input vector for the linear transformation $\mathbf{F}_{ij}$, we have $\mathbf{F}_{ij}\mathbf{x} = ((\mathbf{R}\mathbf{e}_i)\otimes(\mathbf{R}\mathbf{e}_j))\mathbf{x} = (\mathbf{R}\mathbf{e}_j, \mathbf{x})\,\mathbf{R}\mathbf{e}_i = (\mathbf{e}_j, \mathbf{R}^T\mathbf{x})\,\mathbf{R}\mathbf{e}_i = \mathbf{R}((\mathbf{e}_j, \mathbf{R}^T\mathbf{x})\,\mathbf{e}_i) = \mathbf{R}(\mathbf{e}_i\otimes\mathbf{e}_j)(\mathbf{R}^T\mathbf{x}) = \mathbf{R}\mathbf{E}_{ij}\mathbf{R}^T\mathbf{x}$. We have therefore shown that, when (4.19) is the relation between the orthonormal bases **e** and **f** for R, then the corresponding relation between the induced bases **E** and **F** for L is

$$\mathbf{F}_{ij} = \mathbf{R}\,\mathbf{E}_{ij}\,\mathbf{R}^{T}, \tag{4.22}$$

or

$$\mathbf{F}_{ij} = \underline{\underline{\mathbf{R}}}\,\mathbf{E}_{ij}, \tag{4.23}$$

where $\underline{\underline{\mathbf{R}}}$ is the 4-rotation generated by the proper orthogonal 2-tensor **R**. Note the elegant analogy with (4.19).

Now let $\underline{\underline{\mathbf{C}}}$ be any 4-tensor, and let $\mathbf{F}_{ij}$ and $\mathbf{E}_{ij}$ be the two induced bases in L described above, corresponding respectively to the bases **f** and **e** in R. What is the relation between $c^{\mathbf{f}}_{ijkl}$ and $c^{\mathbf{e}}_{ijkl}$? On the one hand, by definition of the components of a 4-tensor and because of (4.23), $\underline{\underline{\mathbf{C}}}\,\mathbf{F}_{kl} = c^{\mathbf{f}}_{ijkl}\mathbf{F}_{ij} = c^{\mathbf{f}}_{ijkl}\,\underline{\underline{\mathbf{R}}}\,\mathbf{E}_{ij}$. On the other hand, by (4.23), $\underline{\underline{\mathbf{C}}}\,\mathbf{F}_{kl} = \underline{\underline{\mathbf{D}}}\,\mathbf{E}_{kl}$, where the 4-tensor $\underline{\underline{\mathbf{D}}}$ is given by $\underline{\underline{\mathbf{D}}} = \underline{\underline{\mathbf{C}}}\,\underline{\underline{\mathbf{R}}}$. Thus $c^{\mathbf{f}}_{ijkl}\,\underline{\underline{\mathbf{R}}}\,\mathbf{E}_{ij} = \underline{\underline{\mathbf{D}}}\,\mathbf{E}_{kl}$. In terms of the respective components $r^{\mathbf{e}}_{ijkl}$ and $d^{\mathbf{e}}_{ijkl}$ of the 4-tensors $\underline{\underline{\mathbf{R}}}$ and $\underline{\underline{\mathbf{D}}}$ in the basis **e**, we thus have $c^{\mathbf{f}}_{ijkl}\,r^{\mathbf{e}}_{pqij}\,\mathbf{E}_{pq} = d^{\mathbf{e}}_{pqkl}\,\mathbf{E}_{pq}$, from which we conclude that $c^{\mathbf{f}}_{ijkl}\,r^{\mathbf{e}}_{pqij} = d^{\mathbf{e}}_{pqkl}$. But since $\underline{\underline{\mathbf{D}}} = \underline{\underline{\mathbf{C}}}\,\underline{\underline{\mathbf{R}}}$, we have $d^{\mathbf{e}}_{pqkl} = c^{\mathbf{e}}_{pqms}\,r^{\mathbf{e}}_{mskl}$. Thus $c^{\mathbf{f}}_{ijkl}\,r^{\mathbf{e}}_{pqij} = c^{\mathbf{e}}_{pqms}\,r^{\mathbf{e}}_{mskl}$. Finally, we make use of the fact that the orthogonal 4-tensor $\underline{\underline{\mathbf{R}}}$ is a 4-rotation; its components are therefore given by (4.18) as $r^{\mathbf{e}}_{ijkl} = r^{\mathbf{e}}_{ik}\,r^{\mathbf{e}}_{jl}$. So we have the relation

$$r^{\mathbf{e}}_{pi}\,r^{\mathbf{e}}_{qj}\,c^{\mathbf{f}}_{ijkl} = r^{\mathbf{e}}_{mk}\,r^{\mathbf{e}}_{sl}\,c^{\mathbf{e}}_{pqms}. \tag{4.24}$$

But the $r^{\mathbf{e}}_{ij}$s, being the components of an orthogonal 2-tensor, satisfy a relation of the form (3.26); using this fact in (4.24) and doing a bit of cosmetic work on the subscripts supplies the *change-of-basis formula* for the components of a 4-tensor:

$$c^{\mathbf{f}}_{ijkl} = r^{\mathbf{e}}_{pi}\,r^{\mathbf{e}}_{qj}\,r^{\mathbf{e}}_{rk}\,r^{\mathbf{e}}_{sl}\,c^{\mathbf{e}}_{pqrs}. \tag{4.25}$$

It should be noted with appreciation that the summation convention is in force in (4.25).

A different approach to the study of Cartesian tensors makes fundamental use of the change-of-basis formula in the very *definition* of a tensor. According to this view, which is that taken, for example, in the book by Spain [4.3], a Cartesian 2-tensor is a rule that assigns to each orthonormal basis a set of $n^2$ real numbers, with the proviso that the assigned numbers in two different bases must conform to the change-of-basis formula (3.40). Similarly, a 4-tensor is a rule that assigns $n^4$ components to each basis, subject to the transformation rule (4.25). This explains the terminology *2-tensor* and *4-tensor*, at least superficially: there are $n^2$ components of a 2-tensor, they are indexed by two subscripts, and they satisfy (3.40). There are $n^4$ components of a 4-tensor, they are indexed by four subscripts, and they satisfy

(4.25). According to this scheme, one would guess that a *3-tensor* is the rule that assigns $n^3$ real numbers $c^{\mathbf{e}}_{ijk}$ to each basis $\mathbf{e}$, subject to a change-of-basis rule $c^{\mathbf{f}}_{ijk} = r^{\mathbf{e}}_{pi} r^{\mathbf{e}}_{qj} r^{\mathbf{e}}_{rk} c^{\mathbf{e}}_{pqr}$. On the other hand, it is not so clear how one would define a 3-tensor in the spirit of the approach to tensors taken in this book. We leave it to the reader to speculate on this issue.

## References

[4.1] M.E. Gurtin, The linear theory of elasticity, in *Handbuch der Physik*, Volume VIa/2, Springer-Verlag, Berlin, 1972.

[4.2] M.E. Gurtin, *An Introduction to Continuum Mechanics*, Academic Press, New York, 1981.

[4.3] B. Spain, *Tensor Calculus*, Oliver and Boyd, Ltd., Edinburgh, 1953.

[4.4] L.C. Martins and P. Podio-Guidugli, A variational approach to the polar decomposition theorem, *Rend. Accad. Naz. Lincei (Classe Sci. Fis., Mat. Natur.)*, Volume 66, pp. 487–93, 1979.

## Problems

**4.1.** Suppose the scalar product of two 2-tensors $\mathbf{A}, \mathbf{B}$ is defined by $\langle \mathbf{A}, \mathbf{B} \rangle = \mathrm{Tr}\,(\mathbf{A}\mathbf{B}^T)$. Show that the induced basis in L is orthonormal: $\langle \mathbf{E}_{ij}, \mathbf{E}_{kl} \rangle = \delta_{ik}\delta_{jl}$.

**4.2.** Let S be the linear manifold in L consisting of all *symmetric* 2-tensors on the n-dimensional real Euclidean space R. Let $\mathbf{e}_1, \ldots \mathbf{e}_n$ be an orthonormal basis for R, with $\mathbf{E}_{ij} = \mathbf{e}_i \otimes \mathbf{e}_j$ the corresponding induced orthonormal basis in L. Set $\mathbf{F}_{ij} = (1/2)(\mathbf{E}_{ij} + \mathbf{E}_{ji})$. Show that the six $\mathbf{F}$s form a basis for the six-dimensional space S. Is this basis orthonormal?

**4.3.** Let $\mathbf{A}$ be a fixed non-singular 2-tensor in L, and define a real-valued function $\delta$ on the set O of all orthogonal 2-tensors $\mathbf{Q}$ by $\delta(\mathbf{Q}) = \|\mathbf{A} - \mathbf{Q}\|^2$. (a) Show that for any $\mathbf{Q}$ in O,

$$\delta(\mathbf{Q}) = \|\mathbf{1}\|^2 + \|\mathbf{A}\|^2 - 2\,\mathrm{Tr}\,(\mathbf{Q}\mathbf{A}^T),$$

where $\mathbf{1}$ is the identity 2-tensor. (b) Use the left polar decomposition $\mathbf{A} = \mathbf{V}\mathbf{R}$, where $\mathbf{V}$ is symmetric and positive definite and $\mathbf{R}$ is orthogonal, to write

$$\delta(\mathbf{Q}) = \|\mathbf{1}\|^2 + \|\mathbf{A}\|^2 - 2\,\mathrm{Tr}\,(\mathbf{Q}\mathbf{R}^T\mathbf{V}) \quad (*);$$

and thus to conclude that

$$\min_{\mathbf{Q}\in\mathsf{O}} \delta(\mathbf{Q}) = \|\mathbf{1}\|^2 + \|\mathbf{A}\|^2 - 2 \max_{\mathbf{P}\in\mathsf{O}} \Delta(\mathbf{P}),$$

where $\Delta(\mathbf{P}) = \mathrm{Tr}\,(\mathbf{PV})$ for all $\mathbf{P}$ in $\mathsf{O}$. (c) Let $p_{ij}$ be the components of $\mathbf{P}$ in a principal basis for $\mathbf{V}$, and let $\lambda_1, \ldots, \lambda_n$ be the eigenvalues of $\mathbf{V}$. Use the result of Problem 3.20 to show that $\Delta(\mathbf{P}) \leq \sum_{i=1}^{n} \lambda_i = \mathrm{Tr}\,\mathbf{V} = \Delta(\mathbf{1})$. (d) Show that min $\delta(\mathbf{Q}) = \delta(\mathbf{R})$, and $\delta(\mathbf{R}) = \|\mathbf{1}\|^2 + \|\mathbf{A}\|^2 - 2\,\mathrm{Tr}\,\{(\mathbf{AA}^T)^{1/2}\}$. This shows that the best *orthogonal* approximation (in the sense of minimum $\delta$) to a given non-singular tensor $\mathbf{A}$ is provided by the orthogonal factor $\mathbf{R}$ in the polar decomposition of $\mathbf{A}$. For further discussion, see the paper by L.C. Martins and P. Podio-Guidugli [4.4] cited in the references for this chapter.

**4.4.** The following question arises in a theory of the so-called *martensitic transformation* in crystals. Let $\mathbf{A}$ and $\mathbf{B}$ be fixed symmetric, positive definite 2-tensors. For any pair of proper orthogonal 2-tensors $\mathbf{P}$ and $\mathbf{Q}$, define $\delta(\mathbf{P}, \mathbf{Q}) = \|\mathbf{PA} - \mathbf{QB}\|^2$. What is the minimum value of $\delta(\mathbf{P}, \mathbf{Q})$ over all proper orthogonal tensors $\mathbf{P}$, $\mathbf{Q}$? By using an argument similar to that used in the preceding problem, show that this minimum is given by min $\delta(\mathbf{P}, \mathbf{Q}) = \|\mathbf{A}\|^2 + \|\mathbf{B}\|^2 - 2\,\mathrm{Tr}\{(\mathbf{BA}^2\mathbf{B})^{1/2}\}$. Simplify this result for the special case in which $\mathbf{A}$ and $\mathbf{B}$ commute.

**4.5.** Suppose a 4-tensor has the major symmetry and one minor symmetry. Show that it also has the other minor symmetry.

**4.6.** Let $\mathbf{Q}$ and $\mathbf{R}$ be fixed orthogonal 2-tensors, and define a 4-tensor $\underline{\underline{\mathbf{P}}}$ by $\underline{\underline{\mathbf{P}}}\mathbf{A} = \mathbf{QAR}$ for every 2-tensor $\mathbf{A}$. Show that $\underline{\underline{\mathbf{P}}}$ is an orthogonal 4-tensor, thus proving that not all orthogonal 4-tensors are 4-rotations.

**4.7.** Let $\mathbf{M}$ and $\mathbf{N}$ be fixed symmetric 2-tensors, and define a 4-tensor $\underline{\underline{\mathbf{L}}}$ by $\underline{\underline{\mathbf{L}}}\mathbf{A} = \mathbf{MAN}$ for every 2-tensor $\mathbf{A}$. Does $\underline{\underline{\mathbf{L}}}$ possess the major symmetry? Either minor symmetry?

**4.8.** What are the components of the 4-tensor product $\mathbf{B}\otimes\mathbf{C}$ of the 2-tensors $\mathbf{B}$ and $\mathbf{C}$? In particular, what are the components of $\mathbf{1}\otimes\mathbf{1}$, where $\mathbf{1}$ is the identity 2-tensor?

**4.9.** Let $\underline{\underline{\mathbf{C}}}$ and $\underline{\underline{\mathbf{D}}}$ be 4-tensors, and set $\underline{\underline{\mathbf{G}}} = \underline{\underline{\mathbf{C}}}\,\underline{\underline{\mathbf{D}}}$. Find the components $g_{ijkl}$ of $\underline{\underline{\mathbf{G}}}$ in an orthonormal basis $\mathbf{e}$.

**4.10.** Let the dimension of the underlying real Euclidean space $\mathsf{R}$ be 3. Let $\underline{\underline{\mathbf{C}}}$ be a 4-tensor with the major symmetry and both minor symmetries, and let $c_{ijkl}$ be its components in a given orthonormal basis. In the absence of *any* symmetries, there are $3^4 = 81$ independently assignable components $c_{ijkl}$. How many are there if $\underline{\underline{\mathbf{C}}}$ has *all three* symmetries?

**4.11.** Show that the components of the transposition 4-tensor $\underline{\underline{\mathbf{T}}}$ are the same in all orthonormal bases $\mathbf{e}$. A tensor with this property is said to be *isotropic*. The notion of isotropy is discussed in the following chapter.

**4.12.** If $c_{ijkl}$ are the components of a 4-tensor in an orthonormal basis, show that $c_{iikk}$ is the same in all orthonormal bases.

**4.13.** Suppose the dimension of R is two, and let $\underline{\underline{\mathbf{T}}}$ be the transposition 4-tensor; recall that $\underline{\underline{\mathbf{T}}}$ is symmetric. Find four linearly independent "eigentensors" **A** of $\underline{\underline{\mathbf{T}}}$ and the corresponding eigenvalues $\alpha$: $\underline{\underline{\mathbf{T}}}\,\mathbf{A} = \alpha\,\mathbf{A}$.

**4.14.** Suppose the dimension of R is two, and let $\underline{\underline{\mathbf{C}}}$ be the 4-tensor $\mathbf{A}\otimes\mathbf{A}$, where **A** is a non-null symmetric 2-tensor. Find the eigenvalues, the eigentensors, and the spectral representation of $\underline{\underline{\mathbf{C}}}$.

**4.15.** Let $\underline{\underline{\mathbf{C}}} = \alpha\,\underline{\underline{\mathbf{1}}} + \beta\,\mathbf{1}\otimes\mathbf{1}$, where $\alpha$, $\beta$ are scalars, $\underline{\underline{\mathbf{1}}}$ is the identity 4-tensor, and **1** is the identity 2-tensor. Find all eigentensors and eigenvalues of the 4-tensor $\underline{\underline{\mathbf{C}}}$; i.e., find all scalars $\lambda$ and the corresponding non-null 2-tensors **A** such that $\underline{\underline{\mathbf{C}}}\,\mathbf{A} = \lambda\mathbf{A}$.

**4.16.** Let $\underline{\underline{\mathbf{C}}}$ be a 4-tensor with the second minor symmetry. Is there a non-null 2-tensor **A** such that $\underline{\underline{\mathbf{C}}}\,\mathbf{A} = \mathbf{O}$? As a linear transformation of L into itself, is $\underline{\underline{\mathbf{C}}}$ singular?

**4.17.** *Positive definite and strongly elliptic 4-tensors.* Let $\underline{\underline{\mathbf{C}}}$ be a 4-tensor with all three symmetries. (a) $\underline{\underline{\mathbf{C}}}$ is said to be *positive-definite* if the quadratic form $\langle\underline{\underline{\mathbf{C}}}\,\mathbf{A},\ \mathbf{A}\rangle$ is positive for all non-null *symmetric* 2-tensors **A**. Write out this quadratic form in terms of the components of $\underline{\underline{\mathbf{C}}}$ and **A**. What is $\langle\underline{\underline{\mathbf{C}}}\mathbf{A},\ \mathbf{A}\rangle$ if $\underline{\underline{\mathbf{C}}} = \underline{\underline{\mathbf{1}}}$? If $\underline{\underline{\mathbf{C}}} = \mathbf{1}\otimes\mathbf{1}$? Are these two special 4-tensors positive definite? (b) $\underline{\underline{\mathbf{C}}}$ is said to be *strongly elliptic* if $\langle\underline{\underline{\mathbf{C}}}\mathbf{B},\ \mathbf{B}\rangle > 0$ for all 2-tensors of the special form $\mathbf{B} = \mathbf{u}\otimes\mathbf{v}$, where **u** and **v** are arbitrary non-null vectors in R. Write this condition out in terms of components. Is $\underline{\underline{\mathbf{C}}} = \underline{\underline{\mathbf{1}}}$ strongly elliptic? $\underline{\underline{\mathbf{C}}} = \mathbf{1}\otimes\mathbf{1}$? Show that, if $\underline{\underline{\mathbf{C}}}$ is positive definite, it is also strongly elliptic, but not conversely. The notion of strong ellipticity originates in the theory of partial differential equations; for a connection between this notion and a physical issue, see the following problem. Strong ellipticity also plays a major role in the continuum modeling of solids that can undergo *phase transitions*, or abrupt changes in crystal structure, when subject to stress.

**4.18.** *The acoustic tensor.* An elastic solid is characterized by a 4-tensor $\underline{\underline{\mathbf{C}}}$ called the *elasticity tensor*; $\underline{\underline{\mathbf{C}}}$ describes the particular elastic material under consideration, and it always has the major symmetry and both minor symmetries. Let the components of $\underline{\underline{\mathbf{C}}}$ in a given orthonormal basis be $c_{ijkl}$. Let **n** be a unit vector with components $n_i$, and let **Q** be the 2-tensor whose components are $q_{ik} = c_{ijkl}\ n_j n_l$. It can be shown that the square of the speed of propagation of a small-amplitude wave traveling in the direction **n** in the elastic medium must be proportional to an eigenvalue of **Q**; the proportionality factor is the reciprocal of the mass density of the material, and is thus positive. (a) Show that **Q** is a symmetric 2-tensor. (b) Show that, for a given **n**, the quadratic form $(\mathbf{m}, \mathbf{Qm}) > 0$ for every unit vector **m** if and only if $c_{ijkl}\, n_j n_l m_i m_k > 0$ for the given unit vector **n** and all unit vectors **m**. (c) Show that this condition is equivalent to the requirement that $\langle \mathbf{A}, \underline{\underline{\mathbf{C}}}\,\mathbf{A} \rangle > 0$ where $\mathbf{A} = \mathbf{n} \otimes \mathbf{m}$, **n** is the given vector in the propagation direction, and the inequality must hold for all unit vectors **m**. (d) Thus show that small-amplitude waves will propagate in *every* direction **n** if and only if $\underline{\underline{\mathbf{C}}}$ is strongly elliptic; refer to the preceding problem for the definition of this term. The symmetric 2-tensor **Q** is called the *acoustic 2-tensor* of the material.

# CHAPTER 5

# APPLICATIONS

In this chapter, we describe some applications in mechanics of the notions introduced in the earlier parts of the book. The issues to be considered are (a) deformations of a continuum; (b) isotropy and material symmetry in solids; (c) scalar-valued functions of 2-tensors and fundamental scalar invariants; and (d) linear dynamical systems with a finite number of degrees of freedom that do not possess "classical normal modes."

## Deformations of a Continuum

We now consider a 3-dimensional real Euclidean space R, and we identify R with ordinary physical space. Vectors $\mathbf{x}$ in R are regarded as position vectors to points in space. Suppose that a body composed of a continuous medium such as a solid or a fluid occupies a region of space consisting of points whose position vectors comprise a subregion D of R, each "particle" of the body being identified with a point in D.

A *deformation* of the body is a mapping that carries the particle $\mathbf{x}$ in D to a new point in space whose position vector is $\mathbf{y}$. The rule that assigns $\mathbf{y}$ to $\mathbf{x}$ is a vector-valued function, say $\hat{\mathbf{y}}$; thus a deformation is described by

$$\mathbf{y} = \hat{\mathbf{y}}(\mathbf{x}) \text{ for all } \mathbf{x} \text{ in D}. \tag{5.1}$$

Much can be learned about the nature of deformations of a solid by studying the special case of (5.1) in which the mapping $\mathbf{x} \to \mathbf{y}$ is of the form

$$\mathbf{y} = \hat{\mathbf{y}}(\mathbf{x}) = \mathbf{F}\mathbf{x} + \mathbf{c}, \tag{5.2}$$

where $\mathbf{F}$ is a Cartesian 2-tensor and $\mathbf{c}$ is a vector, with both $\mathbf{F}$ and $\mathbf{c}$ independent of $\mathbf{x}$. The mapping $\hat{\mathbf{y}}$ may, if desired, be viewed as extended from D to all of R by requiring the rule (5.2) to hold for *all* vectors $\mathbf{x}$, and not just those that correspond to points in D. For simplicity, we shall take this point of view, so that the body under consideration occupies all of three-dimensional space. Deformations of the special form (5.2) are said to be *homogeneous*.

To be physically reasonable, the mapping $\mathbf{x} \to \mathbf{y}$ must be one-to-one; i.e., each particle of the body is carried to one and only one point in space (particles do not break up), and no two particles are carried to the *same* point (particles do not interpenetrate). This means that the tensor $\mathbf{F}$ must be non-singular. Somewhat deeper physical considerations lead to the conclusion that, at least for the simplest and most common class of deformations, the determinant of $\mathbf{F}$ must not only fail to vanish, it must in fact be positive:

$$\det \mathbf{F} > 0. \tag{5.3}$$

Since $\mathbf{F}$ is non-singular, it has right- and left-polar decompositions as described in Chapter 3: there are unique positive definite, symmetric tensors $\mathbf{U}$ and $\mathbf{V}$, together with a unique orthogonal tensor $\mathbf{Q}$, such that

$$\mathbf{F} = \mathbf{QU} = \mathbf{VQ}. \tag{5.4}$$

Since $\det \mathbf{F} = \det \mathbf{Q} \det \mathbf{U}$ and $\det \mathbf{F}$ and $\det \mathbf{U}$ are both positive, $\det \mathbf{Q}$ must also be positive, so that in fact $\det \mathbf{Q} = 1$. Thus $\mathbf{Q}$ must be proper, corresponding to a rotation without reflection. Moreover, $\mathbf{U}^2 = \mathbf{F}^T\mathbf{F}$ and $\mathbf{V}^2 = \mathbf{F}\mathbf{F}^T$, as demonstrated in the proof of the polar decomposition theorem.

Consider a line segment of particles in the body, and let $\mathbf{x}$ and $\mathbf{x}'$ be the position vectors to the points occupied by the particles at the ends of the segment; see Figure 5.1. Suppose that the particles at $\mathbf{x}$ and $\mathbf{x}'$ are carried to the points $\mathbf{y}$ and $\mathbf{y}'$, respectively, by the deformation. In Problem 5.1, it is shown that the deformation (5.2) carries line segments into line segments. The length of the segment is $|\mathbf{x}' - \mathbf{x}|$; what is the length $|\mathbf{y}' - \mathbf{y}|$ of the deformed image of the segment? This question, which is especially important for the mechanics of solids, makes immediate contact with the polar decomposition of $\mathbf{F}$. To shorten the formulas to come, it is helpful to introduce the vectors

$$\mathbf{p} = \mathbf{x}' - \mathbf{x}, \quad \mathbf{q} = \mathbf{y}' - \mathbf{y}; \tag{5.5}$$

see Figure 5.1. Clearly

$$|\mathbf{q}|^2 = |\mathbf{y}' - \mathbf{y}|^2 = |\mathbf{Fx}' - \mathbf{Fx}|^2 = |\mathbf{Fp}|^2 = (\mathbf{F}\mathrm{p}, \mathbf{Fp}) = (\mathbf{p}, \mathbf{F}^T\mathbf{Fp}) = (\mathbf{p}, \mathbf{U}^2\mathbf{p}) \tag{5.6}$$

Thus the deformed length of the segment depends only on the right stretch tensor $\mathbf{U}$, and not on the rotation tensor $\mathbf{Q}$ or the translation vector $\mathbf{c}$. The *change* in the square of the length of the segment during deformation is

$$|\mathbf{q}|^2 - |\mathbf{p}|^2 = 2\,(\mathbf{p}, \mathbf{Ep}), \tag{5.7}$$

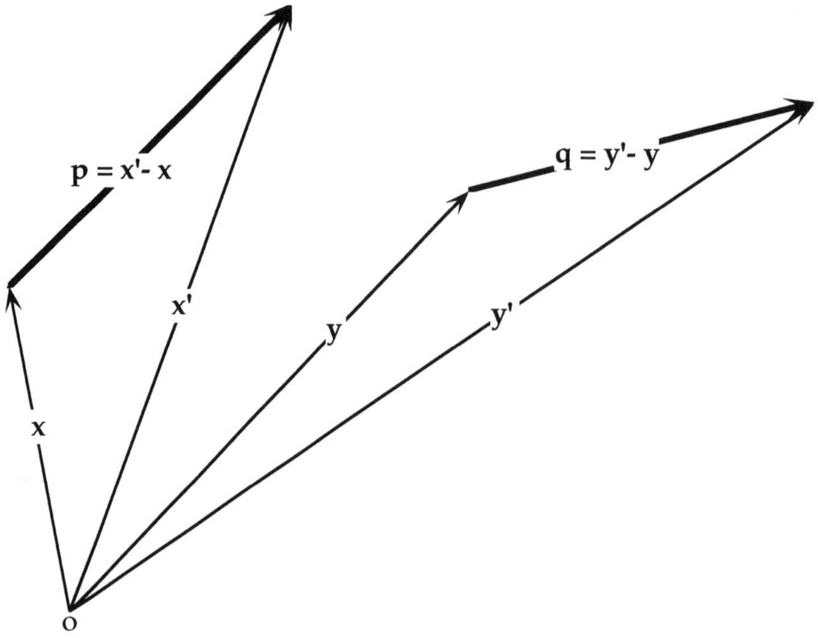


**FIGURE 5.1.** Line segments $\mathbf{p} = \mathbf{x}' - \mathbf{x}$ (before deformation) and $\mathbf{q} = \mathbf{y}' - \mathbf{y}$ (after deformation).

where $\mathbf{E}$ is the *strain tensor*, conventionally defined by

$$\mathbf{E} = (1/2)(\mathbf{U}^2 - \mathbf{1}) = (1/2)(\mathbf{F}^T\mathbf{F} - \mathbf{1}). \tag{5.8}$$

The *relative elongation* $\delta(\mathbf{p})$ of the segment is

$$\delta(\mathbf{p}) \equiv \frac{|\mathbf{q}|}{|\mathbf{p}|} - 1 = \left\{1 + 2\,\frac{(\mathbf{p}, \mathbf{Ep})}{(\mathbf{p}, \mathbf{p})}\right\}^{1/2} - 1. \tag{5.9}$$

It may be observed that, if $\mathbf{p}$ on the right side of (5.9) is replaced by $\alpha\mathbf{p}$, where $\alpha$ is a scalar, the relative elongation is left unchanged; thus in the homogeneous deformation (5.2), the relative elongation of a segment depends only on the *direction* of the undeformed segment, and not on its length.

Since the strain tensor $\mathbf{E}$ is symmetric, it has three orthonormal eigenvectors $\mathbf{e}_1$, $\mathbf{e}_2$, $\mathbf{e}_3$ corresponding to three eigenvalues $\epsilon_1$, $\epsilon_2$, $\epsilon_3$, taken to be ordered so that $\epsilon_1 \leq \epsilon_2 \leq \epsilon_3$. The $\epsilon$s are called the *principal strains* for the deformation. Consider a segment that, in its undeformed state, is parallel to $\mathbf{e}_i$, so that $\mathbf{x}' - \mathbf{x}$ is a scalar multiple of $\mathbf{e}_i$. By (5.9) the relative elongation $\delta(\mathbf{e}_i)$ of such a segment is given by

$$\delta(\mathbf{e}_i) = (1 + 2\epsilon_i)^{1/2} - 1. \tag{5.10}$$

Properties of the quadratic form associated with **E** allow us to show that, among *all* segments, those with direction parallel to $\mathbf{e}_3$, corresponding to the greatest principal strain $\epsilon_3$, suffer the greatest relative elongation, while those parallel to $\mathbf{e}_1$ suffer the least; see Problem 5.3. One should note that these statements refer to the *signed* values of $\delta$, since **E** need not be positive definite. Thus some segments may be contracted by the deformation, while others are lengthened.

Now consider *two* line segments issuing from the common point **x** in the undeformed body, one terminating at the point $\mathbf{x}'$, the other at $\mathbf{x}''$; let $\mathbf{y}$, $\mathbf{y}'$ and $\mathbf{y}''$ be the respective deformed images of the points $\mathbf{x}$, $\mathbf{x}'$, and $\mathbf{x}''$ (Figure 5.2). Remembering (5.5), we let $\mathbf{r} = \mathbf{x}'' - \mathbf{x}$, $\mathbf{s} = \mathbf{y}'' - \mathbf{y}$. Assume that the two segments $\mathbf{p} = \mathbf{x}' - \mathbf{x}$ and $\mathbf{r}$ are perpendicular before deformation, so $(\mathbf{p}, \mathbf{r}) = 0$. The angle $\alpha(\mathbf{p}, \mathbf{r})$ between the deformed images of the segments is determined by the relation $\cos\alpha = (\mathbf{q}, \mathbf{s})/(|\mathbf{q}||\mathbf{s}|)$, or

$$\cos\alpha(\mathbf{p}, \mathbf{r}) = \frac{(\mathbf{Fp}, \mathbf{Fr})}{(\mathbf{Fp}, \mathbf{Fp})^{1/2}\,(\mathbf{Fr}, \mathbf{Fr})^{1/2}}\,. \tag{5.11}$$

The right side of (5.11) is independent of the lengths of **p** and **r**, which may therefore be taken as orthonormal. Rewriting (5.11) and using $\mathbf{F}^T\mathbf{F} = 2\mathbf{E} + \mathbf{1}$ from (5.8) leads to

$$\cos\alpha(\mathbf{p}, \mathbf{r}) = \frac{2\,(\mathbf{p}, \mathbf{Er})}{(1 + 2(\mathbf{p},\mathbf{Ep}))^{1/2}\,(1 + 2(\mathbf{r}, \mathbf{Er}))^{1/2}}\,. \tag{5.12}$$

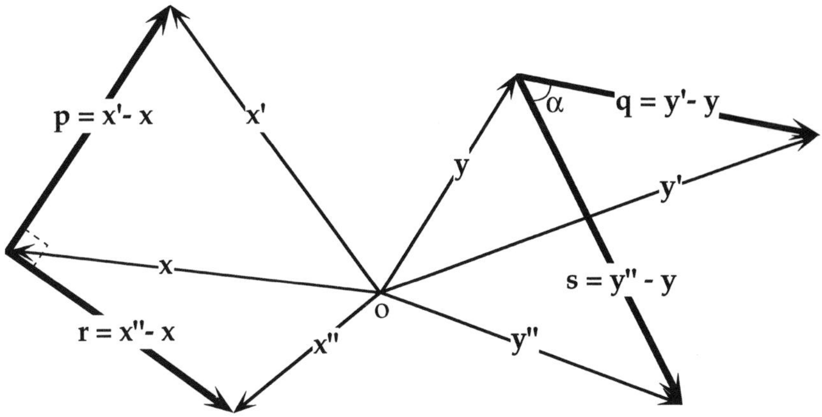


**FIGURE 5.2.** Line segments $\mathbf{p} = \mathbf{x}' - \mathbf{x}$ and $\mathbf{r} = \mathbf{x}'' - \mathbf{x}$ (before deformation) and $\mathbf{q} = \mathbf{y}' - \mathbf{y}$ and $\mathbf{s} = \mathbf{y}'' - \mathbf{y}$ (after deformation).

Thus the angle between the pair of deformed segments is *also* determined by the strain tensor. In particular, suppose that **p** and **r** coincide with a pair of eigenvectors $\mathbf{e}_i$ and $\mathbf{e}_j$ of **E**, respectively. Then $(\mathbf{p}, \mathbf{Er}) = (\mathbf{e}_i, \mathbf{Ee}_j) = \epsilon_j (\mathbf{e}_i, \mathbf{e}_j)$ (no sum on j), so that $(\mathbf{p}, \mathbf{Er}) = 0$. Thus two perpendicular directions that coincide with with eigenvectors of the strain tensor *remain* perpendicular after deformation.

Let $\mathbf{f}_1, \mathbf{f}_2, \mathbf{f}_3$ be an *arbitrary* orthonormal basis for $\mathsf{R}$, and let $E_{ij}$ be the components of the strain tensor in **f**. From (5.9) and (5.12) we conclude that

$$\delta(\mathbf{f}_i) = (1 + 2\,E_{ii})^{1/2} - 1 \quad \text{(no sum on i)}, \tag{5.13}$$

$$\cos\alpha(\mathbf{f}_i, \mathbf{f}_j) = 2\frac{E_{ij}}{(1 + 2E_{ii})^{1/2}\,(1 + 2E_{jj})^{1/2}} \quad \text{(no sum on i or j)}. \tag{5.14}$$

Thus for a segment whose direction before deformation is that of $\mathbf{f}_i$, the relative elongation is determined by the $i^{th}$ diagonal component $E_{ii}$ (no sum) of **E** in **f**, but the angle between the deformed image of a pair of segments in the directions of $\mathbf{f}_i$ and $\mathbf{f}_j$ depends on $E_{ii}$ and $E_{jj}$ (no sums) as well as on $E_{ij}$.

If $\mathbf{U} = \mathbf{1}$, or equivalently $\mathbf{F} = \mathbf{Q}$, then the strain tensor vanishes: $\mathbf{E} = \mathbf{O}$. Conversely, $\mathbf{E} = \mathbf{O}$ implies that $\mathbf{F} = \mathbf{Q}$. Thus during a homogeneous deformation, lengths and angles are preserved if and only if the deformation is of the special form $\mathbf{y} = \mathbf{Qx} + \mathbf{c}$, where **Q** is a proper orthogonal tensor and **c** is a constant vector. Such a deformation is called a *rigid deformation*; it is composed of a rotation **Q** and a translation **c**.

Returning to the general homogeneous deformation (5.2), let $\xi_i$, $\eta_i$, $c_i$, and $F_{ij}$ be the respective components in an orthonormal basis **f** of **x**, **y**, **c**, and **F**, and rewrite (5.2) in component form as $\eta_i = F_{ij}\,\xi_j + c_i$. This determines each $\eta$ as a linear function of the $\xi$s, and the partial derivatives of these functions are given by $\partial\eta_i/\partial\xi_j = F_{ij}$. Thus **F** is appropriately called the *deformation gradient* tensor. Sometimes it is convenient to consider the *displacement* vector $\mathbf{u} = \mathbf{y} - \mathbf{x} = \hat{\mathbf{y}}(\mathbf{x}) - \mathbf{x} = \mathbf{Fx} + \mathbf{c} - \mathbf{x}$; the *displacement gradient* is a 2-tensor denoted by $\nabla\mathbf{u}$ and defined by $\nabla\mathbf{u} = \mathbf{F} - \mathbf{1}$. Its components in the basis **f** are $\partial u_i/\partial\xi_j = F_{ij} - \delta_{ij}$. From $(5.8)_2$, we may write the strain tensor in terms of the displacement gradient as

$$\mathbf{E} = (1/2)[\nabla\mathbf{u} + (\nabla\mathbf{u})^T + (\nabla\mathbf{u})^T\,\nabla\mathbf{u}]. \tag{5.15}$$

The homogeneous deformation (5.2) is said to be *infinitesimal* if $\|\nabla\mathbf{u}\| = \|\mathbf{F} - \mathbf{1}\|$ is small compared to unity. For an infinitesimal deformation, it is natural to neglect the quadratic terms in $\nabla\mathbf{u}$ in (5.15), thus arriving at an approximation $\mathbf{E} \sim \Gamma$, where $\Gamma$ is the so-called *infinitesimal strain tensor*:

$$\Gamma = (1/2)[\nabla\mathbf{u} + (\nabla\mathbf{u})^T]. \tag{5.16}$$

From $(5.8)_1$, we may write the right stretch tensor **U** in the form

$$\mathbf{U} = (\mathbf{1} + 2\mathbf{E})^{1/2}; \tag{5.17}$$

For an infinitesimal deformation, (5.17) formally yields the approximation

$$\mathbf{U} \sim \mathbf{1} + \mathbf{E} \sim \mathbf{1} + \Gamma = \mathbf{1} + (1/2)\,[\nabla\mathbf{u} + (\nabla\mathbf{u})^{\mathrm{T}}], \tag{5.18}$$

while from $(5.4)_1$, $(5.18)_2$, and (5.16),

$$\mathbf{Q} = \mathbf{F}\mathbf{U}^{-1} = (\mathbf{1} + \nabla\mathbf{u})\mathbf{U}^{-1} \sim \mathbf{1} + (1/2)[\nabla\mathbf{u} - (\nabla\mathbf{u})^{\mathrm{T}}]. \tag{5.19}$$

Thus for infinitesimal deformations, the appropriate approximations to the factors $\mathbf{U}$ and $\mathbf{Q}$ of the polar decomposition $(5.4)_1$ are determined by the symmetric part $\Gamma \equiv (1/2)[\nabla\mathbf{u} + (\nabla\mathbf{u})^{\mathrm{T}}]$ and the skew-symmetric part $\Omega \equiv (1/2)[\nabla\mathbf{u} - (\nabla\mathbf{u})^{\mathrm{T}}]$ of the displacement gradient $\nabla\mathbf{u}$: $\mathbf{U} \sim \mathbf{1} + \Gamma$, $\mathbf{Q} \sim \mathbf{1} + \Omega$. When these approximations are returned to $(5.4)_1$ to find the appropriate approximation to $\mathbf{F}$, the result is

$$\mathbf{F} \sim \mathbf{1} + \Gamma + \Omega = \mathbf{1} + \nabla\mathbf{u}; \tag{5.20}$$

but the relation $\mathbf{F} = \mathbf{1} + \nabla\mathbf{u}$ is an *exact* and immediate consequence of the definition of $\nabla\mathbf{u}$!

The theory of infinitesimal deformations constitutes the kinematic basis of the classical linearized theory of elastic solids, a theory that has enjoyed great success when applied to modestly loaded solids composed of a variety of materials. For solids such as rubber capable of undergoing large deformations, the fully nonlinear kinematical theory discussed in the first portion of this section is required in the formulation of appropriate models.

For practical reasons, non-homogeneous deformations—those for which the tensor $\mathbf{F}$, or equivalently $\nabla\mathbf{u}$, is not independent of $\mathbf{x}$—are ultimately the main objects of study in the theory of deformation of solids. But in a small neighborhood of a point $\mathbf{x}$, even a non-homogeneous deformation $\mathbf{x} \rightarrow \mathbf{x} + \mathbf{u}(\mathbf{x})$ is *locally* homogeneous *relative to the point* $\mathbf{x}$. For this reason, the importance of the theory of homogeneous deformations—those of the form (5.2), where $\mathbf{F}$ is constant—is greater than one might have expected.

Finally, whatever happened to the *left* stretch factor $\mathbf{V}$ in the polar decomposition $(5.4)_2$ for $\mathbf{F}$? This mystery is cleared up in Problem 5.2.

## Isotropy and Material Symmetry

In many areas of classical field theory, one ingredient of the model considered is a fundamental relation between two primitive entities in the theory. These entities may be vectors, such as the heat flux and the temperature gradient in the classical theory of heat conduction, or they may be 2-tensors, such as stress and strain in the theory of elasticity. The relation alluded to above that connects the pair of entities under consideration expresses the role played in the theory by a material property—heat conductivity, for example. In the simplest versions of the underlying theories, this relation is a linear one. In the classical theory of heat conduction, for example, the relation is often taken to be that of *Fourier's law*, which asserts that the heat flux vector $\mathbf{q}$ is linear in the temperature gradient $\mathbf{g}$:

$$\mathbf{q} = \mathbf{K}\mathbf{g}. \tag{5.21}$$

Here the *heat conductivity* $\mathbf{K}$ is a symmetric 2-tensor that describes the heat-transfer characteristics of the material. In linear elasticity, the corresponding relation is

$$\Sigma = \underline{\underline{\mathbf{C}}}\, \Gamma, \tag{5.22}$$

where the stress $\Sigma$ and the infinitesimal strain $\Gamma$ are 2-tensors, and the 4-tensor $\underline{\underline{\mathbf{C}}}$, called the *elasticity tensor*, possesses the major symmetry and both minor symmetries.

The tensors $\mathbf{K}$ and $\underline{\underline{\mathbf{C}}}$ are determined by experiments carried out on specimens of the material of interest. In the case of heat conduction, for example, a conceptual version of such an experiment might be described in the following way. From a large square sheet of material, a small square specimen with sides parallel to those of the large sheet is removed for testing; let $\mathbf{e}_1$ and $\mathbf{e}_2$ be unit vectors parallel to the sides of the small square. Next, the specimen is subjected to a given temperature gradient parallel to $\mathbf{e}_1$, and both components of the resulting heat flux vector $\mathbf{q}$ in the basis $\mathbf{e}$ are measured. This allows the determination of the components $k^{\mathbf{e}}_{11}$ and $k^{\mathbf{e}}_{12}$ in the basis $\mathbf{e}$ of the conductivity tensor $\mathbf{K}$. One then subjects the specimen to a given temperature gradient parallel to $\mathbf{e}_2$, again measuring the components of the resulting heat flux vector. This not only determines $k^{\mathbf{e}}_{22}$, but it also delivers $k^{\mathbf{e}}_{21}$, which—by the assumed symmetry of $\mathbf{K}$—should coincide with $k^{\mathbf{e}}_{12}$. We leave aside such important issues as whether the experiment indeed yields $k^{\mathbf{e}}_{12} = k^{\mathbf{e}}_{21}$, or—even more fundamentally—whether the linear relation (5.21) is consistent with experimental results. Instead, we focus on the following question: if the small square specimen had been oriented differently, so that its sides were *not* parallel to the sides of the large sheet, and if $\mathbf{f}_1$ and $\mathbf{f}_2$ were unit vectors parallel to the sides of this reoriented specimen and obtained by rotation from $\mathbf{e}_1$ and $\mathbf{e}_2$, would the experimentally determined values $k^{\mathbf{f}}_{11}$, $k^{\mathbf{f}}_{12}$ and $k^{\mathbf{f}}_{22}$ coincide respectively with $k^{\mathbf{e}}_{11}$, $k^{\mathbf{e}}_{12}$, and $k^{\mathbf{e}}_{22}$? In general, the answer to this question is no. If indeed it happens that the measured components of $\mathbf{K}$ are independent of the orientation of the specimen, so that $k^{\mathbf{f}}_{ij} = k^{\mathbf{e}}_{ij}$ for *any* pair of orthonormal bases $\mathbf{f}$ and $\mathbf{e}$ related through rotation, the material is said to be an *isotropic* heat conductor.

The description above of the conceptual experiment was carried out in a two-dimensional setting for reasons of simplicity, but it is clear that a similar discussion would apply in three dimensions. *From here on in this section of the present chapter, we work in a three-dimensional real Euclidean space* R.

An analogous conceptual experiment can be described for the determination of the elasticity tensor.

As suggested by the discussion of heat conductivity above, we agree to call a 2-tensor **A** (symmetric or not) *isotropic* if its components are the same in all orthonormal bases. The mathematical question is then clear: what is the form of the most general tensor **A** whose components are independent of basis? If a 2-tensor **A** has the same components in two orthonormal bases **e** and **f**, the change-of-basis formula (3.40) shows that the matrix of components $\underline{A}^{e}$ of **A** must satisfy $\underline{A}^{e} = (\underline{R}^{e})^{T}\underline{A}^{e}\underline{R}^{e}$, where $\underline{R}^{e}$ is the orthogonal matrix of components in **e** of the tensor **R** carrying **e** to **f** according to (3.38). If **f** is obtained from **e** by rotation, then **R** is a *proper* orthogonal tensor. It follows immediately that a 2-tensor **A** is isotropic if and only if it satisfies

$$\mathbf{A} = \mathbf{QAQ}^{T} \text{ for every proper orthogonal tensor } \mathbf{Q}. \qquad (5.23)$$

Put differently, **A** is isotropic if and only if it commutes with every proper orthogonal 2-tensor **Q**: $\mathbf{AQ} = \mathbf{QA}$. Still another way of saying (5.23) is $\mathbf{A} = \underline{\underline{\mathbf{Q}}}\mathbf{A}$ for every 4-rotation $\underline{\underline{\mathbf{Q}}}$. Thus an isotropic tensor is unaffected by rotation.

We shall now determine the most general isotropic *symmetric* tensor. The corresponding result for an *arbitrary* 2-tensor is the subject of Problem 5.4. We now establish the following proposition:

---

***Proposition 5.1.*** A symmetric 2-tensor **A** on a 3-dimensional real Euclidean space is isotropic if and only if $\mathbf{A} = \alpha\mathbf{1}$, where $\alpha$ is a scalar.

---

To prove this, we begin by observing that $\mathbf{A} = \alpha\mathbf{1}$ is obviously *sufficient* for the isotropy of **A**, since **1** has the same components in any basis. Alternatively, $\mathbf{A} = \alpha\mathbf{1}$ clearly commutes with every proper orthogonal tensor.

Necessity is slightly more difficult. Let **A** be an isotropic symmetric tensor. Being symmetric, **A** has 3 orthonormal eigenvectors, call them $\mathbf{a}_1$, $\mathbf{a}_2$, $\mathbf{a}_3$, corresponding to the respective eigenvalues $\alpha_1$, $\alpha_2$, $\alpha_3$. By isotropy, for *any* proper orthogonal tensor **Q**, we have $\mathbf{A} = \mathbf{QAQ}^{T}$, so if $\mathbf{Aa}_i = \alpha_i\mathbf{a}_i$ (no sum), then it is also true that $\mathbf{Ab}_i = \alpha_i\mathbf{b}_i$ (no sum), where the vector $\mathbf{b}_i$ is given by $\mathbf{b}_i = \mathbf{Qa}_i$, and **Q** may be chosen to be *any* proper orthogonal tensor. First, let us choose for **Q** the tensor for which $\mathbf{b}_1 = \mathbf{Qa}_1 = \mathbf{a}_3$, $\mathbf{b}_2 = \mathbf{Qa}_2 = \mathbf{a}_2$, $\mathbf{b}_3 = \mathbf{Qa}_3 = -\mathbf{a}_1$. The matrix of components of this **Q** in the basis **a** is given by

$$\underline{Q}^{\mathbf{a}} = \begin{pmatrix} 0 & 0 & -1 \\ 0 & 1 & 0 \\ 1 & 0 & 0 \end{pmatrix}, \tag{5.24}$$

so det $\underline{Q}^{\mathbf{a}} = 1$, and $\mathbf{Q}$ is indeed proper. It follows that $\mathbf{A}\mathbf{a}_1 = \alpha_3 \mathbf{a}_1$, from which one concludes that $\alpha_3 = \alpha_1$. A similar argument shows that $\alpha_2 = \alpha_1$, so that in fact the all the eigenvalues of $\mathbf{A}$ are the same. Let $\alpha = \alpha_1 = \alpha_2 = \alpha_3$ be their common value, so that $\mathbf{A}\alpha_1 = \alpha \mathbf{a}_i$, i = 1, 2, 3. It follows that the matrix $\underline{A}^{\mathbf{a}}$ of $\mathbf{A}$ in the basis $\mathbf{a}$ is given by $\underline{A}^{\mathbf{a}} = \alpha \underline{1}$, so that indeed $\mathbf{A} = \alpha \mathbf{1}$, and the proposition is proved.

The result of the proposition established above and the method of proof remain valid in a real Euclidean space of *any* finite dimension.

The story for isotropic elastic solids is more complicated. As in heat conduction, we say that an elastic material is isotropic if the components of the elasticity 4-tensor are the same in all orthonormal bases $\mathbf{f}$ and $\mathbf{e}$ connected by a proper orthogonal tensor: $c^{\mathbf{f}}_{ijkl} = c^{\mathbf{e}}_{ijkl}$. According to the change-of-basis formula (4.25) of Chapter 4, this means that, for any proper orthogonal tensor $\mathbf{Q}$ with components $q^{\mathbf{e}}_{ij}$ in $\mathbf{e}$, we must have

$$c^{\mathbf{e}}_{ijkl} = c^{\mathbf{f}}_{ijkl} = q^{\mathbf{e}}_{pi} q^{\mathbf{e}}_{qj} q^{\mathbf{e}}_{rk} q^{\mathbf{e}}_{sl}\, c^{\mathbf{e}}_{pqrs}. \tag{5.25}$$

Let Z be the set of all 4-rotations $\underline{\underline{\mathbf{Q}}}$ defined by $\underline{\underline{\mathbf{Q}}}\,\mathbf{A} = \mathbf{Q}\mathbf{A}\mathbf{Q}^{T}$ for any 2-tensor $\mathbf{A}$ in L and any proper orthogonal 2-tensor $\mathbf{Q}$. By (4.17), the relation (5.25) satisfied by the components of the 4-tensor $\underline{\underline{\mathbf{C}}}$ is equivalent to the 4-tensor statement

$$\underline{\underline{\mathbf{C}}} = \underline{\underline{\mathbf{Q}}}\,\underline{\underline{\mathbf{C}}}\,\underline{\underline{\mathbf{Q}}}^{T} \text{ for every } \underline{\underline{\mathbf{Q}}} \text{ in Z}. \tag{5.26}$$

As shown in Problem 4.6, the collection Z of all 4-rotations does *not coincide* with the set of *all* orthogonal 4-tensors. If it did, the proposition about isotropic 2-tensors established above for spaces of arbitrary finite dimension could be used to assert immediately that $\underline{\underline{\mathbf{C}}}$ is a scalar multiple of the identity 4-tensor. This is not the case.

Since the elasticity tensor $\underline{\underline{\mathbf{C}}}$ has the major and both minor symmetries, its restriction to the linear manifold S of *symmetric* tensors in L is a symmetric linear transformation of the six-dimensional real Euclidean space S into itself. Since in elasticity the tensors of interest as inputs to $\underline{\underline{\mathbf{C}}}$ are always symmetric, it is sufficient to consider only this restriction of $\underline{\underline{\mathbf{C}}}$ to S. In all that follows in this section of Chapter 5, we shall use the symbol $\underline{\underline{\mathbf{C}}}$ to stand for this restriction, and we shall work entirely within the space S. Thus for example, when the identity 4-tensor appears in the following discussion, it is understood to be the restriction to S of the identity 4-tensor on L.

---

***Proposition 5.2.*** $\underline{\underline{\mathbf{C}}}$ is the restriction to $\mathsf{S}$ of an isotropic elasticity 4-tensor with the major symmetry and both minor symmetries if and only if

$$\underline{\underline{\mathbf{C}}} = 2\mu\, \underline{\underline{\mathbf{1}}} + \lambda\, \mathbf{1}\otimes\mathbf{1}, \qquad (5.27)$$

where $\mu$ and $\lambda$ are scalars, and $\underline{\underline{\mathbf{1}}}$ is the restriction to $\mathsf{S}$ of the identity 4-tensor.

---

The product $\mathbf{1}\otimes\mathbf{1}$ in (5.27) is that for 2-tensors; see Example 4.2.

Note that if $\underline{\underline{\mathbf{C}}}$ has the form (5.27), the relation (5.22) between stress $\Sigma$ and strain $\Gamma$ becomes

$$\Sigma = 2\mu\, \Gamma + \lambda\, (\mathrm{Tr}\, \Gamma)\, \mathbf{1}; \qquad (5.28)$$

in terms of the components $\sigma_{ij}$ of $\Sigma$ and $\gamma_{ij}$ of $\Gamma$ in any orthonormal basis, (5.28) asserts that

$$\sigma_{ij} = 2\mu\, \gamma_{ij} + \lambda\, \gamma_{kk}\, \delta_{ij}. \qquad (5.29)$$

This is the relation between the components of stress and the components of strain that lies at the heart of the classical theory of elasticity for infinitesimal deformations of an isotropic solid.

To prove the last proposition, we begin by demonstrating that (5.27) is sufficient for isotropy. Indeed, by Problem 5.8, the 4-tensor identity $\underline{\underline{\mathbf{1}}}$ and the 4-tensor product $\mathbf{1}\otimes\mathbf{1}$ of the 2-tensor identity with itself are both isotropic 4-tensors; since a linear combination of two isotropic tensors is obviously isotropic, the 4-tensor $\underline{\underline{\mathbf{C}}}$ given by (5.27) is clearly isotropic.

To show that (5.27) is necessary for isotropy, we begin by observing that, because $\underline{\underline{\mathbf{C}}}$ is a symmetric linear transformation of $\mathsf{S}$ into itself, there are six eigenvalues $\alpha_1, \ldots, \alpha_6$ of $\underline{\underline{\mathbf{C}}}$ and six orthonormal symmetric 2-tensors $\mathbf{A}_1, \ldots, \mathbf{A}_6$ that are the corresponding "eigentensors" of $\underline{\underline{\mathbf{C}}}$:

$$\underline{\underline{\mathbf{C}}}\, \mathbf{A}_i = \alpha_i\, \mathbf{A}_i, \quad \text{(no sum on i)},\ i = 1, \ldots 6, \qquad (5.30)$$

$$\langle \mathbf{A}_i, \mathbf{A}_j \rangle = \delta_{ij}. \qquad (5.31)$$

The spectral formula for the 4-tensor $\underline{\underline{\mathbf{C}}}$ then furnishes the representation

$$\underline{\underline{\mathbf{C}}} = \sum_{i=1}^{6} \alpha_i\, \mathbf{A}_i\otimes\mathbf{A}_i. \qquad (5.32)$$

Next, we show that the identity 2-tensor $\mathbf{1}$ is an eigentensor of $\underline{\underline{\mathbf{C}}}$. By the isotropy of $\underline{\underline{\mathbf{C}}}$,

$$\underline{\underline{\mathbf{C}}}\ \mathbf{1} = \underline{\underline{\mathbf{Q}}}\,\underline{\underline{\mathbf{C}}}\,\underline{\underline{\mathbf{Q}}}^{\mathrm{T}}\ \mathbf{1} = \underline{\underline{\mathbf{Q}}}\,\underline{\underline{\mathbf{C}}}(\mathbf{Q}\mathbf{1}\mathbf{Q}^{\mathrm{T}}) = \underline{\underline{\mathbf{Q}}}\,\underline{\underline{\mathbf{C}}}\ \mathbf{1} = \mathbf{Q}(\underline{\underline{\mathbf{C}}}\ \mathbf{1})\mathbf{Q}^{\mathrm{T}}, \qquad (5.33)$$

for every proper orthogonal 2-tensor $\mathbf{Q}$. Thus $\underline{\underline{\mathbf{C}}}\mathbf{1}$ is a symmetric isotropic 2-tensor, so that by Proposition 5.1 above,

$$\underline{\underline{\mathbf{C}}}\mathbf{1} = \alpha_1 \mathbf{1} \tag{5.34}$$

for some scalar $\alpha_1$. In other words, $\mathbf{A}_1 \equiv 3^{-1/2}\,\mathbf{1}$ is an eigentensor of unit length for $\underline{\underline{\mathbf{C}}}$ corresponding to the eigenvalue $\alpha_1$. There are five remaining unit eigentensors $\mathbf{A}_2, ..., \mathbf{A}_6$ of $\underline{\underline{\mathbf{C}}}$ in $\mathsf{S}$; since each of these is also orthogonal to $\mathbf{A}_1$, we have

$$\langle \mathbf{A}_i, \mathbf{1} \rangle = \mathrm{Tr}\, \mathbf{A}_i = 0,\ i = 2, ..., 6. \tag{5.35}$$

Thus $\mathbf{A}_2, ..., \mathbf{A}_6$ comprise an orthonormal basis for the 5-dimensional linear manifold $\mathsf{T}$ in $\mathsf{S}$ consisting of all *traceless* symmetric tensors.

We now prove

---

***Proposition 5.3.*** Let $\mathbf{A}$ be a symmetric, traceless eigentensor of $\underline{\underline{\mathbf{C}}}$ corresponding to the eigenvalue $\alpha$: $\underline{\underline{\mathbf{C}}}\,\mathbf{A} = \alpha\mathbf{A}$. Let $\mathbf{Q}$ be any proper orthogonal 2-tensor, and set $\mathbf{B} = \mathbf{Q}\mathbf{A}\mathbf{Q}^{\mathrm{T}}$. Then $\mathbf{B}$ is also a symmetric, traceless eigentensor of $\underline{\underline{\mathbf{C}}}$ corresponding to the eigenvalue $\alpha$; $\underline{\underline{\mathbf{C}}}\,\mathbf{B} = \alpha\mathbf{B}$.

---

It is clear that $\mathbf{B}$ is symmetric and traceless. Let $\underline{\underline{\mathbf{Q}}}$ be the 4-rotation associated with $\mathbf{Q}$, so that $\mathbf{B} = \underline{\underline{\mathbf{Q}}}\mathbf{A}$. Then because $\underline{\underline{\mathbf{C}}}$ is isotropic, $\underline{\underline{\mathbf{C}}}\,\underline{\underline{\mathbf{Q}}} = \underline{\underline{\mathbf{Q}}}\,\underline{\underline{\mathbf{C}}}$, and thus $\underline{\underline{\mathbf{C}}}\,\mathbf{B} = \underline{\underline{\mathbf{C}}}\,\underline{\underline{\mathbf{Q}}}\,\mathbf{A} = \underline{\underline{\mathbf{Q}}}\,\underline{\underline{\mathbf{C}}}\,\mathbf{A} = \underline{\underline{\mathbf{Q}}}\,(\alpha\mathbf{A}) = \alpha\,\underline{\underline{\mathbf{Q}}}\,\mathbf{A} = \alpha\mathbf{B}$, so that $\mathbf{B}$ is indeed an eigentensor of $\underline{\underline{\mathbf{C}}}$ corresponding to the eigenvalue $\alpha$, establishing the proposition.

Now consider the eigentensor $\mathbf{A}_2$ of $\underline{\underline{\mathbf{C}}}$; since $\mathbf{A}_2$ is symmetric, traceless and of unit length, there is an orthonormal basis $\mathbf{e}_1, \mathbf{e}_2, \mathbf{e}_3$ for $\mathsf{R}$ in which the matrix $\mathrm{A}_2$ of components of $\mathbf{A}_2$ has the form

$$\mathrm{A}_2 = (2\beta^2 + 2\gamma^2 + 2\beta\gamma)^{-1/2}\begin{pmatrix} \beta & 0 & 0 \\ 0 & \gamma & 0 \\ 0 & 0 & -(\beta+\gamma) \end{pmatrix}, \tag{5.36}$$

where $\beta$ and $\gamma$ are scalars. One can easily show that there is a proper orthogonal 2-tensor $\mathbf{Q}$ such that the matrix of components of $\mathbf{A}'_2 \equiv \mathbf{Q}\mathbf{A}_2\mathbf{Q}^{\mathrm{T}}$ is that obtained by interchanging $\beta$ and $\gamma$ in (5.36). By Proposition 1, $\mathbf{A}'_2$ is also an eigentensor of $\underline{\underline{\mathbf{C}}}$ corresponding to the eigenvalue $\alpha_2$, and $\mathbf{A}_2 + \mathbf{A}'_2$ therefore has this property as well. We therefore lose no generality in assuming that $\beta = \gamma$, and hence that

$$\mathrm{A}_2 = 6^{-1/2}\begin{pmatrix} 1 & 0 & 0 \\ 0 & 1 & 0 \\ 0 & 0 & -2 \end{pmatrix}. \tag{5.37}$$

***Proposition 5.4.*** There are four proper orthogonal 2-tensors $\mathbf{Q}_1$, $\mathbf{Q}_2$, $\mathbf{Q}_3$, $\mathbf{Q}_4$ such that the four traceless, symmetric tensors $\mathbf{B}_i$ defined through

$$\mathbf{B}_i = \mathbf{Q}_i\mathbf{A}_2\mathbf{Q}_i^T, \quad i = 1, 2, 3, 4, \tag{5.38}$$

together with $\mathbf{A}_2$ itself, comprise a basis for the linear manifold $\mathsf{T}$ of traceless symmetric 2-tensors in $\mathsf{S}$.

To show this, we begin by choosing $\mathbf{Q}_1$ and $\mathbf{Q}_2$ to be rotations about the eigenvector $\mathbf{e}_1$ of $\mathbf{A}_2$ through the angles $\pi/4$ and $-\pi/4$, respectively. This leads through (5.38) to tensors $\mathbf{B}_1$ and $\mathbf{B}_2$ whose matrices of components in $\mathbf{e}$ are

$$B_1 = 6^{-1/2}\begin{pmatrix} 1 & 0 & 0 \\ 0 & -1/2 & -3/2 \\ 0 & -3/2 & -1/2 \end{pmatrix}, \quad B_1 = 6^{-1/2}\begin{pmatrix} 1 & 0 & 0 \\ 0 & -1/2 & 3/2 \\ 0 & 3/2 & -1/2 \end{pmatrix}. \tag{5.39}$$

Similarly, choosing $\mathbf{Q}_3$ to be the rotation about $\mathbf{e}_2$ through the angle $\pi/4$ gives

$$B_3 = 6^{-1/2}\begin{pmatrix} -1/2 & 0 & 3/2 \\ 0 & 1 & 0 \\ 3/2 & 0 & -1 \end{pmatrix}. \tag{5.40}$$

Finally, let $\mathbf{Q}_4$ be the rotation through $\pi/4$ about $\mathbf{e}_1 - \mathbf{e}_2$; this gives

$$B_4 = 6^{-1/2}\begin{pmatrix} 1/4 & 3/4 & -3(8)^{-1/2} \\ 3/4 & 1/4 & 3(8)^{-1/2} \\ -3(8)^{-1/2} & -3(8)^{-1/2} & -1/2 \end{pmatrix}. \tag{5.41}$$

Direct calculation shows that $\mathbf{A}_2$ and the four $\mathbf{B}_i$s are linearly independent, and therefore form a basis, although not an orthonormal one, for $\mathsf{T}$. This demonstrates Proposition 5.4.

By Proposition 5.3, $\mathbf{A}_2$ and the four $\mathbf{B}_i$s have the common eigenvalue $\alpha_2$. It follows that *every* tensor in $\mathsf{T}$, being a linear combination of these five tensors, is an eigentensor of $\underline{\underline{\mathbf{C}}}$ with eigenvalue $\alpha_2$; in particular, this is true for $\mathbf{A}_i$, $i = 2, \ldots, 6$. The spectral representation (5.32) thus becomes

$$\underline{\underline{\mathbf{C}}} = \alpha_1\mathbf{A}_1\otimes\mathbf{A}_1 + \alpha_2\sum_{i=2}^{6}\mathbf{A}_i\otimes\mathbf{A}_i. \tag{5.42}$$

But applying the spectral formula to the restriction to $\mathsf{S}$ of the identity 4-tensor gives

$$\underline{\underline{\mathbf{1}}} = \sum_{i=1}^{6}\mathbf{A}_i\otimes\mathbf{A}_i, \tag{5.43}$$

so that (5.43) may be rewritten as

$$\underline{\underline{\mathbf{C}}} = (\alpha_1 - \alpha_2)\, \mathbf{A}_1 \otimes \mathbf{A}_1 + \alpha_2\, \underline{\underline{\mathbf{1}}} \tag{5.44}$$

Bearing in mind that $\mathbf{A}_1 = 3^{-1/2}\, \mathbf{1}$, we set $\lambda = (\alpha_1 - \alpha_2)/3$, $\mu = \alpha_2/2$ in (5.44), thus establishing the representation (5.27) for the restriction to $\mathsf{S}$ of an isotropic 4-tensor possessing all symmetries.

Isotropy is one form of *material symmetry* that an elastic solid may exhibit. A more general look at the notion of material symmetry is facilitated by observing first that the set $\mathsf{O}^+$ of all proper orthogonal 2-tensors forms a *group* under the operation of multiplication of tensors: the inverse of every tensor in $\mathsf{O}^+$ is also in $\mathsf{O}^+$, as is the product of any pair of tensors in $\mathsf{O}^+$. Moreover, the group $\mathsf{O}^+$ contains *subgroups*: subsets of $\mathsf{O}^+$ that are themselves groups under the multiplicative operation. In imposing the restriction of isotropy on the elasticity tensor $\underline{\underline{\mathbf{C}}}$, we require that (5.26) should hold for 4-rotations generated by *all* proper orthogonal 2-tensors in $\mathsf{O}^+$. A less demanding restriction would impose (5.26) only for all proper orthogonal 2-tensors in a specified subgroup of $\mathsf{O}^+$. Pursuit of this program leads to the elucidation of the various crystal classes, corresponding to the various subgroups. The isotropic crystals are those possessing the highest degree of material symmetry.

## Scalar Functions of Tensors and Fundamental Scalar Invariants

Let $\mathsf{R}$ be an n-dimensional real Euclidean space, and let $\mathsf{L}$ be the set of all linear transformations of $\mathsf{R}$ into itself. Let $(.,.)$ and $\langle .,. \rangle$ be the respective scalar products on $\mathsf{R}$ and $\mathsf{L}$. Here we shall be concerned with scalar-valued functions $\varphi$ on $\mathsf{L}$ or perhaps on subsets of $\mathsf{L}$. In the case $n = 3$, for example, the *stored elastic energy per unit volume* associated with the homogeneous deformation (5.2) is a scalar-valued function $\varphi(\mathbf{F})$ of the deformation gradient tensor $\mathbf{F}$; in this case, (5.3) shows that the real-valued function $\varphi$ need be defined only on the set $\mathsf{L}^+$ of all 2-tensors $\mathbf{F}$ with positive determinant.

Returning to the n-dimensional case, let us consider a scalar-valued function $\varphi$ on the space $\mathsf{L}$: thus $\varphi(\mathbf{A})$ is a real number for every 2-tensor $\mathbf{A}$ in $\mathsf{L}$. Let $\mathbf{e}_1, ..., \mathbf{e}_n$ be an orthonormal basis for $\mathsf{R}$. Since a tensor is determined by a knowledge of its components in $\mathbf{e}$, there is a real-valued function $\Phi^{\mathbf{e}}$ on the set $L$ of all $n \times n$ matrices such that

$$\varphi(\mathbf{A}) = \Phi^{\mathbf{e}}(\underline{A}^{\mathbf{e}}) \text{ for every } \mathbf{A} \text{ in } \mathsf{L} \tag{5.45}$$

where $\underline{A}^{\mathbf{e}}$ is the matrix of components of $\mathbf{A}$ in $\mathbf{e}$. We begin by finding a change-of-basis formula for $\Phi^{\mathbf{e}}$. Let $\mathbf{f}_1, ..., \mathbf{f}_n$ be a second orthonormal basis for R, with $\mathbf{f}_i = \mathbf{R}\mathbf{e}_i$ for some proper orthogonal tensor $\mathbf{R}$. Then $\varphi(\mathbf{A}) = \Phi^{\mathbf{f}}(\underline{A}^{\mathbf{f}}) = \Phi^{\mathbf{e}}(\underline{A}^{\mathbf{e}})$, so that, by the change-of-basis formula (3.40) for 2-tensors, $\varphi(\mathbf{A}) = \Phi^{\mathbf{f}}(\underline{R}^{\mathbf{e}T}\underline{A}^{\mathbf{e}}\underline{R}^{\mathbf{e}}) = \Phi^{\mathbf{e}}(\underline{A}^{\mathbf{e}})$. It follows that the value $\Phi^{\mathbf{f}}(\underline{A})$ of $\Phi^{\mathbf{f}}$ at any matrix $\underline{A}$ in $L$ is given in terms of $\Phi^{\mathbf{e}}$ by

$$\Phi^{\mathbf{f}}(\underline{A}) = \Phi^{\mathbf{e}}(\underline{R}^{\mathbf{e}}\underline{A}\underline{R}^{\mathbf{e}T}) \text{ for any A in } L. \tag{5.46}$$

The scalar-valued function $\varphi$ on L is said to be an *isotropic* scalar function, or a *scalar invariant*, if

$$\varphi(\underline{\underline{\mathbf{Q}}}\mathbf{A}) = \varphi(\mathbf{A}) \text{ for every 2-tensor } \mathbf{A} \text{ in L and every 4-rotation } \underline{\underline{\mathbf{Q}}}; \tag{5.47}$$

equivalently,

$$\varphi(\mathbf{Q}\mathbf{A}\mathbf{Q}^{T}) = \varphi(\mathbf{A}) \text{ for every } \mathbf{A} \text{ in L and every } \mathbf{Q} \text{ in } \mathsf{O}^{+}. \tag{5.48}$$

Thus rotation of the argument 2-tensor $\mathbf{A}$ does not affect the value of the function $\varphi$. Examples of isotropic scalar functions are furnished by $\varphi(\mathbf{A}) = \mathrm{Tr}\,\mathbf{A}$ and $\varphi(\mathbf{A}) = \det \mathbf{A}$.

For an isotropic scalar function, it follows immediately from (5.48) that, for any orthonormal basis $\mathbf{e}$,

$$\Phi^{\mathbf{e}}(\underline{Q}\underline{A}\underline{Q}^{T}) = \Phi^{\mathbf{e}}(\underline{A}) \text{ for any } \underline{A} \text{ in } L \tag{5.49}$$

and for any proper orthogonal matrix $\underline{Q}$. Using this result with $\underline{Q} = \underline{R}^{\mathbf{e}}$ in (5.45) provides the important conclusion that, for an *isotropic* scalar function $\varphi$,

$$\Phi^{\mathbf{f}}(\underline{A}) = \Phi^{\mathbf{e}}(\underline{A}) \text{ for every } \underline{A} \text{ in } L \tag{5.50}$$

and for every pair of orthonormal bases $\mathbf{e}$, $\mathbf{f}$. Thus in the isotropic case, $\varphi(\mathbf{A}) = \Phi(\underline{A}^{\mathbf{e}})$, where $\Phi$ is independent of basis.

From here on, we limit our attention to the practically important special case of scalar functions $\varphi$ whose domain is the set S of all *symmetric* tensors in L. Equations (5.45)–(5.50) remain valid in this case provided L and $L$ are replaced in the qualifying conditions in these equations by S and $S$, respectively; here $S$ is the set of all symmetric $n \times n$ matrices. Supposing $\varphi$ to be isotropic, one has

$$\varphi(\mathbf{A}) = \Phi(\underline{A}^{\mathbf{e}}) \text{ for every } \mathbf{A} \text{ in S}, \tag{5.51}$$

for any orthonormal basis $\mathbf{e}$. Since $\Phi$ in (5.51) is independent of basis and $\mathbf{A}$ is symmetric, we may choose for $\mathbf{e}$ a principal basis for $\mathbf{A}$, in which case (5.51) reduces immediately to

$$\varphi(\mathbf{A}) = \hat{\Phi}(\lambda_1, \lambda_2, \ldots, \lambda_n) \text{ for every } \mathbf{A} \text{ in S}, \tag{5.52}$$

where $\lambda_1, \ldots, \lambda_n$ are the eigenvalues of $\mathbf{A}$. In (5.49), choose $\underline{Q}$ to be the proper orthogonal matrix that coincides with the identity except for the $2 \times 2$ block in the upper left corner, which is $\begin{pmatrix} 0 & -1 \\ 1 & 0 \end{pmatrix}$, and choose for $\underline{A}$ the diagonal matrix with entries $\lambda_1, ..., \lambda_n$. The result shows that $\hat{\Phi}(\lambda_1, \lambda_2, \lambda_3, ..., \lambda_n) = \hat{\Phi}(\lambda_2, \lambda_1, \lambda_3 ..., \lambda_n)$. Similarly, one shows that $\lambda_i$ and $\lambda_j$ may be interchanged in the argument of $\hat{\Phi}$ for *any* i and j, so that $\hat{\Phi}$ is a *symmetric* function of the $\lambda_i$s. Thus an isotropic scalar function on S is a symmetric function of the eigenvalues of **A** that does not depend on the eigenvectors of **A**.

Now let us consider the special case n = 2. Then the relation (5.52) for an isotropic scalar function $\varphi$ becomes $\varphi(\mathbf{A}) = \hat{\Phi}(\lambda_1, \lambda_2)$. Since $\lambda_1$ and $\lambda_2$ can be interchanged in the argument of $\Phi\hat{}$, there is no loss of generality in assuming that $\lambda_1 \leq \lambda_2$. Set

$$I_1(\mathbf{A}) = \mathrm{Tr}\,\mathbf{A}, \quad I_2(\mathbf{A}) = \det \mathbf{A} \quad \text{for every } \mathbf{A} \text{ in } \mathsf{S}. \tag{5.53}$$

Then also

$$I_1 = \lambda_1 + \lambda_2, \quad I_2 = \lambda_1\lambda_2; \tag{5.54}$$

$I_1$ and $I_2$ satisfy $I_2 \leq I_1{}^2/4$. The mapping (5.54) from the region $\lambda_1 \leq \lambda_2$ in the $\lambda_1$, $\lambda_2$-plane to the region $I_2 \leq I_1{}^2/4$ in the $I_1$, $I_2$-plane is one-to-one, so that one can write the relation $\varphi(\mathbf{A}) = \hat{\Phi}(\lambda_1, \lambda_2)$ in the equivalent form

$$\varphi(\mathbf{A}) = \psi(I_1(\mathbf{A}), I_2(\mathbf{A})) \tag{5.55}$$

for a suitable function $\psi$ defined on the region $I_2 \leq I_1{}^2/4$ in the $I_1$, $I_2$-plane. Thus in the two-dimensional case, an isotropic scalar function of the positive definite symmetric tensor **A** is always expressible in terms of $I_1(\mathbf{A})$ and $I_2(\mathbf{A})$. For this reason, $I_1(\mathbf{A})$ and $I_2(\mathbf{A})$—which are themselves isotropic scalar functions of **A**—are called the *fundamental scalar invariants* of **A**.

There is a precisely analogous story in a three-dimensional real Euclidean space, where it can be shown that an isotropic scalar function $\varphi$ may be represented in the form

$$\varphi(\mathbf{A}) = \psi(I_1(\mathbf{A}), I_2(\mathbf{A}), I_3(\mathbf{A})), \tag{5.56}$$

where the three fundamental scalar invariants $I_1$, $I_2$, $I_3$ are defined by

$$\left.\begin{aligned} I_1(\mathbf{A}) &= \lambda_1 + \lambda_2 + \lambda_3 = \mathrm{Tr}\,\mathbf{A}, \\ I_2(\mathbf{A}) &= \lambda_1\lambda_2 + \lambda_2\lambda_3 + \lambda_3\lambda_1 = (1/2)[(\mathrm{Tr}\,\mathbf{A})^2 - \mathrm{Tr}\,(\mathbf{A}^2)], \\ I_3(\mathbf{A}) &= \lambda_1\lambda_2\lambda_3 = \det \mathbf{A}. \end{aligned}\right\} \tag{5.57}$$

The domain in $I_1$, $I_2$, $I_3$-space on which the function $\psi$ of (5.56) is defined is substantially more complicated than its counterpart in two dimensions.

## Linear Dynamical Systems

A dynamic system with n degrees of freedom is often described by a differential equation of the form

$$\mathbf{M}\ddot{\mathbf{y}} + \mathbf{D}\dot{\mathbf{y}} + \mathbf{K}\mathbf{y} = \mathbf{g}. \tag{5.58}$$

Here $\mathbf{y}(t)$ and $\mathbf{g}(t)$ are vectors in an n-dimensional real Euclidean space $\mathsf{R}$; $\mathbf{y}(t)$ describes the state of the system at time t, and $\mathbf{g}$ is the external excitation vector. The over-dot in (5.58) indicates differentiation with respect to t. The tensors $\mathbf{M}$, $\mathbf{D}$, and $\mathbf{K}$ are symmetric and independent of the time; $\mathbf{M}$ is positive definite. If the system is a mechanical structure, $\mathbf{M}$, $\mathbf{D}$, and $\mathbf{K}$ represent the mass, the damping and the stiffness of the system, respectively, and $\mathbf{K}$ and $\mathbf{D}$ are then positive definite as well.

Let $\mathbf{N}$ be the unique symmetric, positive definite square root of $\mathbf{M}$, and set $\mathbf{y} = \mathbf{N}^{-1}\mathbf{x}$ in (5.58), which then becomes

$$\ddot{\mathbf{x}} + \mathbf{A}\dot{\mathbf{x}} + \mathbf{B}\mathbf{x} = \mathbf{f}, \tag{5.59}$$

where $\mathbf{A} = \mathbf{N}^{-1}\mathbf{D}\mathbf{N}^{-1}$, $\mathbf{B} = \mathbf{N}^{-1}\mathbf{K}\mathbf{N}^{-1}$, and $\mathbf{f} = \mathbf{N}^{-1}\mathbf{g}$. We shall consider the differential equation in the form (5.59).

According to Proposition 3.4, $\mathbf{A}$ and $\mathbf{B}$ share a common principal basis if and only if they commute: $\mathbf{AB} = \mathbf{BA}$. Assume for the moment that this is the case, and let $x_1, \ldots, x_n$ and $f_1, \ldots, f_n$ be the components of $\mathbf{x}$ and $\mathbf{f}$ in a basis that is principal for both $\mathbf{A}$ and $\mathbf{B}$. Referring (5.59) to this basis yields n *uncoupled* scalar differential equations for the $x_i$s:

$$\ddot{x}_i + \alpha_i\dot{x}_i + \beta_i x_i = f_i \quad \text{(no sum on i)}, \ i = 1, \ldots, n. \tag{5.60}$$

where the $\alpha_i$s and $\beta_i$s are the eigenvalues of $\mathbf{A}$ and $\mathbf{B}$, respectively. Thus when $\mathbf{A}$ and $\mathbf{B}$ satisfy the so-called *convenience hypothesis* $\mathbf{AB} = \mathbf{BA}$, the problem of solving (5.59) subject to appropriate initial conditions is reduced to that of solving n uncoupled scalar second-order initial value problems. In this circumstance, the dynamic system is said to possess *classical normal modes*. The study of systems with this property was apparently begun by Lord Rayleigh [5.4] and continued by others, notably Caughey and O'Kelly [5.5].

When $\mathbf{A}$ and $\mathbf{B}$ do *not* commute, dealing with (5.59) is more difficult. An approximation scheme that is sometimes employed in the structural mechanics literature when $\mathbf{AB} \neq \mathbf{BA}$ proceeds as follows: refer (5.59) to a basis that is principal for $\mathbf{B}$, and then *discard* the off-diagonal terms in the resulting matrix of components of $\mathbf{A}$ in this basis. This again reduces the problem to n uncoupled scalar differential equations, but now their solution provides only an approximation to that of the original problem. Moreover, the circumstances under which this approximation is a good one are unclear.

To illuminate the approximation scheme just described, we ask the following question: among all symmetric 2-tensors **T** that commute with **B**, which one provides the best approximation to **A** in the sense of minimizing $\|\mathbf{A} - \mathbf{T}\| = \{\mathrm{Tr}\,(\mathbf{A} - \mathbf{T})^2\}^{1/2}$? In answering this question, we shall consider only the case in which the eigenvalues of **B** are distinct. In this case, **B** has a unique principal basis **e**, and—since **T** must commute with **B**—this basis will be principal for **T** as well. Let $a_{ij}$ be the components of **A** in the basis **e**, and let $\tau_1, ..., \tau_n$ be the eigenvalues of **T**. Then

$$\|\mathbf{A} - \mathbf{P}\|^2 = \sum_{i=1}^{n} \sum_{j=1}^{n} (a_{ij} - \tau_i\delta_{ij})^2 = \sum\sum_{i \neq j}^{n\ n} a^2_{ij} + \sum_{i=1}^{n} (a_{ii} - \tau_i)^2. \quad (5.61)$$

Thus the best choice for the $\tau_i$s is clearly $\tau_i = a_{ii}$, $i = 1, ..., n$. The matrix of components in the basis **e** of the optimal **T** is therefore diagonal with diagonal entries $a_{11}, ..., a_{nn}$. This is precisely the matrix obtained by referring **A** to the principal basis for **B**, and then discarding the resulting off-diagonal terms in the matrix **A** in this basis. Thus the approximation scheme described in the preceding paragraph leads to the best approximation to **A** among all symmetric tensors that commute with **B**, at least when the eigenvalues of **B** are distinct.

If **A** is positive definite, necessarily $a_{11}, ..., a_{nn}$, and therefore $\tau_1, ..., \tau_n$, are all positive (Problem 3.30), so the optimal **T** is positive definite in this case as well.

The approximate approach to (5.59) discussed above treats **A** and **B** unsymmetrically. While there are often good practical reasons for approximating the damping rather than the stiffness, the following question nevertheless suggests itself: among all pairs of symmetric tensors **S**, **T** such that **ST** = **TS**, which pair **S**, **T** minimizes $\|\mathbf{A} - \mathbf{S}\|^2 + \|\mathbf{B} - \mathbf{T}\|^2$? If one can find such an optimizing pair **S**,**T**, one would then replace **A** by **S** and **B** by **T** in (5.59), leading to an approximating system that possess classical normal modes. This interesting question is studied in the paper [5.6] listed below.

## References


[5.1] M.E. Gurtin, The linear theory of elasticity, in *Handbuch der Physik*, Volume VIa/2, Springer-Verlag, Berlin, 1972.

[5.2] M.E. Gurtin, *An Introduction to Continuum Mechanics*, Academic Press, New York, 1981.

[5.3] P. Chadwick, *Continuum Mechanics: Concise Theory and Problems*, John Wiley and Sons, New York, 1976.

[5.4] Lord Rayleigh, *Theory of Sound*, Volume 1, Dover Publications, New York, 1945.

[5.5] T.K. Caughey and M.E.J. O'Kelly, Classical normal modes in damped linear systems, *Journal of Applied Mechanics*, Volume 32, pp. 583–8, 1965.

[5.6] J.K. Knowles, On a minimization problem associated with linear dynamical systems, *Linear Algebra and its Applications*, Volume 165, pp. 173–84, 1992.

## Problems

**5.1.** Suppose that $L$ is a line segment in the region $\mathsf{D}$. Show that the image of $L$ under the deformation (5.2) is also a line segment. Establish the analogous result for planes in $\mathsf{D}$.

**5.2.** To explain the role of the *left* stretch tensor $\mathbf{V}$ of the polar decomposition (5.4), consider a line segment lying in the image domain $\mathsf{D}'$ of $\mathsf{D}$ under the mapping (5.2). Calculate the relative elongation of such a segment under the deformation $\mathbf{x} = \hat{\mathbf{x}}(\mathbf{y}) = \mathbf{F}^{-1}\mathbf{y}$ *inverse* to (5.2). The strain tensor for this deformation is called the *Eulerian* strain tensor, to contrast it with the *Lagrangian* strain tensor $\mathbf{E}$ of (5.8). Express the Eulerian strain tensor in terms of $\mathbf{V}$.

**5.3.** Show that the line segments in $\mathsf{D}$ that suffer the greatest (or least) relative elongation under the deformation (5.2) are those directed along $\mathbf{e}_3$ (or $\mathbf{e}_1$), where $\mathbf{e}_3$ (or $\mathbf{e}_1$) is the eigenvector of $\mathbf{E}$ associated with the largest (or smallest) principal strain $\epsilon_3$ (or $\epsilon_1$).

**5.4.** Suppose the real Euclidean space $\mathsf{R}$ has dimension three, and let $\mathbf{A}$ be an *arbitrary* 2-tensor, not necessarily symmetric. Suppose that $\mathbf{A}$ is isotropic, so that $\mathbf{QAQ}^{\mathrm{T}} = \mathbf{A}$ for every proper orthogonal tensor $\mathbf{Q}$. (a) Let $\mathbf{A} = \mathbf{S} + \boldsymbol{\Omega}$, where $\mathbf{S}$ is symmetric and $\boldsymbol{\Omega}$ is skew-symmetric. Show that $\mathbf{S}$ and $\boldsymbol{\Omega}$ are both isotropic 2-tensors. (b) Show that the only skew-symmetric 2-tensor $\boldsymbol{\Omega}$ is the null tensor $\mathbf{O}$. (c) Thus show that the isotropic tensor $\mathbf{A}$ must be a scalar multiple of the identity, even though $\mathbf{A}$ need not be symmetric.

**5.5.** Let $\mathsf{R}$ be a *two*-dimensional real Euclidean space, and use (5.23) as the definition of isotropy of the tensor $\mathbf{A}$. (a) Show that every skew-symmetric tensor is isotropic. (b) What is the most general isotropic tensor on a two-dimensional real Euclidean space? (c) Suppose that *all* orthogonal tensors $\mathbf{Q}$ are admitted in (5.23), instead of only the proper ones. Show that the only isotropic skew-symmetric tensor is the null tensor. What is the most general isotropic tensor on a two-dimensional space when all orthogonal tensors are admitted in (5.23)?

**5.6.** Define an *isotropic vector* $\mathbf{x}$ in an n-dimensional real Euclidean space $\mathsf{R}$ as a vector whose components are the same in all orthonormal bases. Show that the only isotropic vector is $\mathbf{x} = \mathbf{o}$.

**5.7.** In the 2-dimensional real Euclidean space $\mathsf{R}_2$ of columns of two real numbers $x_1$, $x_2$ with the natural scalar product, consider a deformation of the form described by (2.4), where $\mathbf{A}$ is the *simple shear with amount of shear* k given by

$$\mathbf{A} = \begin{pmatrix} 1 & k \\ 0 & 1 \end{pmatrix};$$

in the interest of simple calculation, take $k = 8/3$. (a) Interpret the deformation geometrically by determining the deformed image of the unit square $\mathsf{D}$: $0 \le x_1 \le 1$, $0 \le x_2 \le 1$. (b) Find the respective right and left polar decompositions $\mathbf{A} = \mathbf{RU}$ and $\mathbf{A} = \mathbf{VR}$ of $\mathbf{A}$. (c) Decompose the deformation into two steps, corresponding to the right polar decomposition: $\mathbf{x}' = \mathbf{Ux}$, $\mathbf{y} = \mathbf{Rx}'$, and carefully sketch the geometric effects of these two steps (the stretch $\mathbf{U}$ followed by the rigid body rotation $\mathbf{R}$) on the square $\mathsf{D}$. (d) Do the same for the left polar decomposition of $\mathbf{A}$.

**5.8.** Show that the identity 4-tensor $\underline{\underline{\mathbf{C}}} = \underline{\underline{\mathbf{1}}}$ and the 4-tensor product $\underline{\underline{\mathbf{C}}} = \mathbf{1} \otimes \mathbf{1}$ of the 2-tensor identity with itself are both isotropic 4-tensors by showing that (5.26) holds for these $\underline{\underline{\mathbf{C}}}$s.

**5.9.** Let $\underline{\underline{\mathbf{C}}}$ be the restriction to $\mathsf{S}$ of an isotropic 4-tensor with all symmetries, so that (5.27) holds. Find necessary conditions on $\lambda$ and $\mu$ for $\underline{\underline{\mathbf{C}}}$ to be positive definite. Find the corresponding restrictions for strong ellipticity of $\underline{\underline{\mathbf{C}}}$; see Problem 4.17.

**5.10.** Let $\varphi$ be a real-valued function on the set $\mathsf{L}$ of cartesian tensors on an n-dimensional real Euclidean space $\mathsf{R}$. Let $\mathbf{e}_1, ..., \mathbf{e}_n$ be an orthonormal basis for $\mathsf{R}$, and represent $\varphi$ in the form $\varphi(\mathbf{A}) = \Phi^e(a^{\mathbf{e}}_{11}, a^{\mathbf{e}}_{12}, ..., a^{\mathbf{e}}_{nn})$, where $\Phi^e$ is a real-valued function defined and continuously differentiable for all values of its arguments. Let $\mathbf{B}$ be the tensor whose components in $\mathbf{e}$ are $b^{\mathbf{e}}_{ij} = \partial\Phi^e(a^{\mathbf{e}}_{11}, a^{\mathbf{e}}_{12}, ..., a^{\mathbf{e}}_{nn})/\partial a^{\mathbf{e}}_{ij}$. (a) Show that the components of $\mathbf{B}$ in a second orthonormal basis $\mathbf{f}$ are given by $b^{\mathbf{f}}_{ij} = \partial\Phi^f(a^{\mathbf{f}}_{11}, a^{\mathbf{f}}_{12}, ..., a^{\mathbf{f}}_{nn})/\partial a^{\mathbf{f}}_{ij}$. The tensor $\mathbf{B}$ is called the *gradient* of $\varphi$ at $\mathbf{A}$ and is commonly denoted by $\mathbf{B} = \varphi_{\mathbf{A}}(\mathbf{A})$. (b) Assuming $\Phi^e$ to be *twice* continuously differentiable, introduce the 4-tensor $\underline{\underline{\mathbf{C}}}$ by specifying its components to be given by $c^{\mathbf{e}}_{ijkl} = \partial^2\Phi^e(a^{\mathbf{e}}_{11}, a^{\mathbf{e}}_{12}, ..., a^{\mathbf{e}}_{nn})/\partial a^{\mathbf{e}}_{ij}\,\partial a^{\mathbf{e}}_{kl}$. Establish for $\underline{\underline{\mathbf{C}}}$ the result analogous to that for $\mathbf{B}$ in part (a). One writes $\underline{\underline{\mathbf{C}}} = \varphi_{\mathbf{AA}}(\mathbf{A})$ and speaks of $\underline{\underline{\mathbf{C}}}$ as the *second* gradient of $\varphi$ at $\mathbf{A}$.

**5.11.** If the dimension of $\mathsf{R}$ is $n = 2$, use the results of Problem 5.10 to show that the gradient of the fundamental scalar invariant $I_1(\mathbf{A})$ at $\mathbf{A}$ is the identity 2-tensor $\mathbf{1}$, and the gradient of $I_2(\mathbf{A})$ at $\mathbf{A}$ is $(\mathbf{1}\otimes\mathbf{1})\mathbf{A} - \mathbf{A}^T$.

**5.12.** Let $\varphi$ be an isotropic scalar function on the set $\mathsf{L}$ of all cartesian tensors on a two-dimensional real Euclidean space $\mathsf{R}$. Under appropriate conditions of differentiability, use the result of Problem 5.11 to show that (5.55) implies that $\varphi_{\mathbf{A}}(\mathbf{A}) = (\partial\psi/\partial I_1)\,\mathbf{1} + (\partial\psi/\partial I_2)[(\mathbf{1}\otimes\mathbf{1})\mathbf{A} - \mathbf{A}^T]$, where the derivatives of $\psi$ are evaluated at $I_1(\mathbf{A})$, $I_2(\mathbf{A})$.

**5.13.** Let $\underline{\underline{\mathbf{C}}}$ be an isotropic 4-tensor with the major and both minor symmetries. Set $\varphi(\mathbf{A}) = \langle \mathbf{A}, \underline{\underline{\mathbf{C}}}\mathbf{A}\rangle$ for every symmetric two tensor $\mathbf{A}$. Show that $\varphi$ is an isotropic scalar-valued function on $\mathsf{S}$. When $\underline{\underline{\mathbf{C}}}$ is the elasticity tensor and $\mathbf{A} = \Gamma$ is the infinitesimal strain tensor, $(1/2)\varphi(\Gamma)$ represents the stored elastic energy per unit volume according to the theory of infinitesimal deformations.

**5.14.** Let $\mathsf{R}$ be a two-dimensional real Euclidean space, and let $\mathbf{A}$ and $\mathbf{B}$ be symmetric tensors whose matrices of components in a given orthonormal basis are given by

$$\underline{A} = \begin{pmatrix} 1 & 4 \\ 4 & 3 \end{pmatrix}, \quad \underline{B} = \begin{pmatrix} 0 & 2 \\ 2 & 1 \end{pmatrix}.$$

Does the dynamical system represented by the differential equation (5.59) have classical normal modes in this case?

**5.15.** Let $\mathsf{R}$ be a two-dimensional real Euclidean space, and consider the differential equation

$$\ddot{\mathbf{x}} + \mathbf{A}\dot{\mathbf{x}} + \mathbf{B}\mathbf{x} = 0, \quad (*)$$

where $\mathbf{A}$ and $\mathbf{B}$ are symmetric tensors. Let $\mathbf{e}_1$, $\mathbf{e}_2$ be a given orthonormal basis in $\mathsf{R}$, and denote by $\underline{x}^{\mathbf{e}}$ the column $\binom{x_1^{\mathbf{e}}}{x_2^{\mathbf{e}}}$ of components of a vector $\mathbf{x}$ in the basis $\mathbf{e}$. Suppose that the matrices $\underline{A}^{\mathbf{e}}$ and $\underline{B}^{\mathbf{e}}$ of components in $\mathbf{e}$ of $\mathbf{A}$ and $\mathbf{B}$ are given by

$$\underline{A}^{\mathbf{e}} = \begin{pmatrix} 2 & 0 \\ 0 & 6 \end{pmatrix}, \quad \underline{B}^{\mathbf{e}} = \begin{pmatrix} 2 & 1 \\ 1 & 2 \end{pmatrix},$$

so that, when referred to the basis $\mathbf{e}$, (*) becomes

$$\ddot{\underline{x}}^{\mathbf{e}} + \begin{pmatrix} 2 & 0 \\ 0 & 6 \end{pmatrix} \dot{\underline{x}}^{\mathbf{e}} + \begin{pmatrix} 2 & 1 \\ 1 & 2 \end{pmatrix} \underline{x}^{\mathbf{e}} = \underline{0}. \quad (**)$$

(a) Show that vectors $\mathbf{f}_1$, $\mathbf{f}_2$ whose columns of components in $\mathbf{e}$ are $\underline{f}_1^{\mathbf{e}} = 2^{-1/2}\binom{1}{1}$, $\underline{f}_2^{\mathbf{e}} = 2^{-1/2}\binom{-1}{1}$ comprise a principal basis for $\mathbf{B}$, with corresponding eigenvalues $\beta_1 = 3$, $\beta_2 = 1$. (b) Find the matrix $\underline{T}^{\mathbf{e}}$

of components in **e** of the tensor **T** whose matrix of components in **f** is obtained by deleting the off-diagonal terms from the matrix $\underline{A}^{\mathbf{f}}$. Thus show that, according to the scheme described under the heading, Scalar Functions of Tensors and Fundamental Scalar Invariants of the present chapter, the differential equation approximating (*), when referred to the basis **f**, is given by

$$\ddot{\underline{x}}^{\mathbf{f}} + 4\,\dot{\underline{x}}^{\mathbf{f}} + \begin{pmatrix} 3 & 0 \\ 0 & 1 \end{pmatrix} \underline{x}^{\mathbf{f}} = \underline{0}. \quad (***)$$

(c) Solve both (**) and (***) subject to the initial conditions $\underline{x}\,(0) = \binom{1}{0}$, $\dot{\underline{x}}\,(0) = \binom{0}{1}$, and compare the exact and approximate solution *vectors* **x** of the initial value problem; remember that $\underline{x}^{\mathbf{e}}$ and $\underline{x}^{\mathbf{f}}$ refer to components of **x** in different bases.

# APPENDIX 1

# ASSUMED BACKGROUND

Here we record without proof or interpretative comment some results from elementary algebra that are taken for granted in the text. These results pertain to the arithmetic of matrices and determinants, and to systems of linear algebraic equations. For a detailed discussion of the manipulative aspects of the matters discussed below, the reader may consult Chapter 1 of the book by Hildebrand [A.2] cited at the end of this Appendix.

It is assumed that the reader is acquainted with the arithmetic of $n \times n$ matrices $\underline{A}$ of real or complex numbers:

$$\underline{A} \equiv \begin{pmatrix} a_{11} & a_{12} & \dots & a_{1n} \\ \dots & \dots & \dots & \dots \\ a_{n1} & a_{n2} & \dots & a_{nn} \end{pmatrix} \equiv (a_{ik}). \tag{A.1}$$

The identity matrix $\underline{1}$ has elements $\delta_{ik}$, where $\delta_{ik}$ is the Kronecker delta: $\delta_{ik} = 1$ if $i = k$, 0 if $i \neq k$. If $\underline{A} = (a_{ik})$ and $\underline{B} = (b_{ik})$ are two $n \times n$ matrices, their sum $\underline{C} = \underline{A} + \underline{B}$ and product $\underline{D} = \underline{A}\underline{B}$ are the matrices whose elements are given by $c_{ik} = a_{ik} + b_{ik}$ and $d_{ik} = \sum_{j=1}^{n} a_{ij} b_{jk}$, respectively. The product $\alpha\underline{A}$ of a number $\alpha$ with the matrix $\underline{A}$ is the matrix with elements $\alpha a_{ik}$.

Associated with the matrix $\underline{A}$ of (A.1) is a number det $\underline{A}$ called the *determinant* of $\underline{A}$. The determinant of the identity matrix is det $\underline{1} = 1$. For the general matrix $\underline{A}$, the determinant may be calculated in various ways. A recursive scheme for finding det $\underline{A}$ involves the notion of the *cofactor* $A_{ik}$ of an element $a_{ik}$ of $\underline{A}$, defined as follows:

$$A_{ik} = \text{cofactor } a_{ik} = (-1)^{i+k} \cdot \text{(determinant of the } (n-1) \times (n-1) \text{ matrix obtained by deleting from } \underline{A} \text{ the row and column containing } a_{ik}) \tag{A.2}$$

The determinant of $\underline{A}$ is then given by either of the following two formulas:

$$\det \underline{A} = \sum_{k=1}^{n} a_{ik} A_{ik} = \sum_{k=1}^{n} a_{ki} A_{ki}. \tag{A.3}$$

The task of calculating det $\underline{A}$ is thus reduced to that of finding the determinants of $2 \times 2$ matrices. If $n = 2$ in (A.1), then by definition, det $\underline{A} = a_{11}a_{22} - a_{12}a_{21}$. Note that the determinant of $\underline{A}$ is a polynomial of degree n in the elements of $\underline{A}$.

The determinant of the product of two $n \times n$ matrices is the product of their separate determinants. Also, det $(\alpha\underline{A}) = \alpha^n$ det $\underline{A}$.

The *transpose* $\underline{A}^T$ of an $n \times n$ matrix $\underline{A}$ is the matrix obtained from $\underline{A}$ by interchanging its rows and columns; $(\underline{AB})^T = \underline{B}^T\underline{A}^T$, and det $\underline{A}^T$ = det $\underline{A}$.

If det $\underline{A} \neq 0$, then there is a matrix denoted by $\underline{A}^{-1}$ and called the *inverse* of $\underline{A}$ such that $\underline{AA}^{-1} = \underline{A}^{-1}\underline{A} = \underline{1}$; the element $a_{ik}^{-1}$ in the $i^{th}$-row and $k^{th}$-column of $\underline{A}^{-1}$ is given by $a_{ik}^{-1} = (\det \underline{A})^{-1} A_{ki}$. Clearly $\sum_{k=1}^{n} a_{ik}^{-1} a_{kj} = \sum_{k=1}^{n} a_{ik} a_{kj}^{-1} = \delta_{ij}$. Since det $\underline{A}$ det $\underline{A}^{-1}$ = det $\underline{AA}^{-1}$ = det $\underline{1}$ = 1, necessarily det $\underline{A}^{-1}$ = 1/det $\underline{A}$.

Consider now the problem of finding a solution or solutions of n linear equations for n unknowns $x_1, x_2, ..., x_n$:

$$\sum_{k=1}^{n} a_{ik} x_k = b_i, \quad i = 1, ..., n. \tag{A.4}$$

Here the $n^2$ numbers $a_{ik}$ and the n numbers $b_i$, all of which may be either real or complex, are given, and the $x_k$s are to be determined. Let $\underline{A} = (a_{ik})$. We refer to the following list of assertions about the system (A.4) collectively as *Cramer's rule*:

***Case 1.*** Suppose det $\underline{A} \neq 0$. Then (A.4) has one and only solution $x_1, ..., x_n$; in particular, if $b_1 = b_2 = ... = b_n = 0$, then the only solution of (A.4) is $x_1 = x_2 = ... = x_n = 0$.

***Case 2.*** Suppose det $\underline{A} = 0$. Then the homogeneous system (A.4) with $b_1 = b_2 = ... = b_n = 0$ has a solution $x_1, ..., x_n$ in which not all the $x_k$s are zero. When the $b_i$s do not all vanish, the non-homogeneous system (A.4) may or may not have a solution. Whether or not the $b_i$s all vanish, when (A.4) has a solution, it is not unique.

It is to be emphasized that Cramer's rule applies whether the $a_{ik}$s, $b_i$s and $x_k$s are real or complex.

## References

[A.1] R. Bellman, *Introduction to Matrix Analysis*, McGraw-Hill, New York, 1960.

[A.2] F.B. Hildebrand, *Methods of Applied Mathematics*, Prentice-Hall, New York, 1952.

APPENDIX 2

# SOLUTIONS FOR SELECTED PROBLEMS

## Chapter 1

**1.1.** No. Linear combinations of solutions of the new differential equation are not themselves solutions.

**1.2.** Let $p(t) = t^N$, $q(t) = -t^N$. Then $p + q$ is the null polynomial, which is not in the set of all polynomials of degree N.

**1.3.** Let $\mathbf{x}_k$ be the null vector in the set. Choose scalars $\alpha_1 = \alpha_2 = \cdots = \alpha_{k-1} = 0$, $\alpha_k = 1$. Then $\alpha_1\mathbf{x}_1 + \alpha_2\,\mathbf{x}_2 + \cdots + \alpha_k\,\mathbf{x}_k = \mathbf{x}_k = \mathbf{o}$; since not all $\alpha$s are zero, the set of **x**s is linearly dependent.

**1.5.** It may be verified directly that cos t and sin t are linearly independent solutions of (1.2). Let $\varphi_1$, $\varphi_2$, $\varphi_3$ be *any* triplet of solutions of (1.2). Each of these solutions can be represented as a linear combination of cos t and sin t. We can assume that $\varphi_1$ and $\varphi_2$ are linearly independent, for otherwise the triplet of $\varphi$s is *trivially* linearly dependent. One can then show that cos t and sin t are expressible as linear combinations of $\varphi_1(t)$ and $\varphi_2(t)$. It follows that $\varphi_3(t)$ is a linear combination of $\varphi_1(t)$ and $\varphi_2(t)$, and therefore that the triplet of $\varphi$s is a linearly dependent set. This argument makes no use of the theory of differential equations.

**1.6.** Let $\alpha_1, \alpha_2, \ldots, \alpha_k$ be scalars, and set $\Phi(t) = \sum \alpha_j\,\varphi_j(t)$. Then the $i^{th}$ derivative of $\Phi$ at $t = 0$ is given by $\Phi^{(i)}(0) = i!\,\alpha_i$. Suppose the $\alpha$s are such that $\Phi(t) \equiv 0$. Then $\Phi^{(i)}(0) = 0$ for $i = 0, 1, 2, \ldots, k$, so that all the $\alpha$s vanish, proving the linear independence of the set of $\varphi$s. This argument makes use of calculus. Can you find a purely algebraic argument?

**1.7.** Suppose **x** can be represented in two ways in the basis **e**: $\mathbf{x} = \sum \alpha_j\,\mathbf{e}_j =$

$\sum \beta_j \mathbf{e}_j$, where the $\alpha$s and $\beta$s are *two* sets of components of $\mathbf{x}$ in $\mathbf{e}$. Then $\sum (\alpha_j - \beta_j)\, \mathbf{e}_j = \mathbf{o}$. By the linear independence of the $\mathbf{e}$s, one finds that $\alpha_j = \beta_j$ for $j = 1, ..., n$.

**1.9.** (a) Let R be the linear space consisting of the complex numbers, with *real* scalars. Let $\mathbf{e}_1 = 1$ and $\mathbf{e}_2 = i$. Then if $z = x + yi$ is in R, $z = x\,\mathbf{e}_1 + y\,\mathbf{e}_2$. Since the $\mathbf{e}$s are linearly independent, they form a basis, and dim R = 2. (b) Now let R be the space of complex numbers, with *complex* scalars. Let $z_1$ be any *fixed* non-null complex number, and set $\mathbf{e}_1 = z_1$. Let z be any element of R. Then $z = (z/z_1)\,\mathbf{e}_1$. The set consisting of the vector $\mathbf{e}_1$ alone is a linearly independent set and forms a basis for R. Thus dim R = 1.

**1.10.** Any linear combination of two functions in $C^{(1)}$ is also in $C^{(1)}$, and $C^{(1)}$ is a subset of C.

**1.11.** Let $\mathbf{e}_1 = \begin{pmatrix} 0 & 1 & 0 \\ -1 & 0 & 0 \\ 0 & 0 & 0 \end{pmatrix}$, $\mathbf{e}_2 = \begin{pmatrix} 0 & 0 & 1 \\ 0 & 0 & 0 \\ -1 & 0 & 0 \end{pmatrix}$, $\mathbf{e}_3 = \begin{pmatrix} 0 & 0 & 0 \\ 0 & 0 & 0 \\ 0 & 0 & -1 \end{pmatrix}$; these are elements of R′, and they are readily shown to comprise a linearly independent set. Moreover, any matrix in R′ is obviously a linear combination of these three $\mathbf{e}$'s. Thus dim R′ = 3.

**1.12.** One can verify that $(\mathbf{x}, \mathbf{y})^{(1)} = x_1y_1 + x_2y_2$ and $(\mathbf{x}, \mathbf{y})^{(2)} = x_1y_1 + (x_1y_2 + x_2y_1)/2 + x_2y_2$ both satisfy the rules imposed on scalar products. (Be sure to check the requirement that $(\mathbf{x}, \mathbf{x})^{(2)} > 0$.) Let $\mathbf{x} = \binom{1}{0}$, $\mathbf{y} = \binom{0}{1}$. Then $(\mathbf{x}, \mathbf{y})^{(1)} = 0$, so the angle between $\mathbf{x}$ and $\mathbf{y}$ induced by the first scalar product is 90°. Next, $(\mathbf{x}, \mathbf{y})^{(2)} = 1/2$, so the angle in the second case is 30°. Note that $\mathbf{x}$ and $\mathbf{y}$ are unit vectors with respect to *both* scalar products.

**1.13.** Let $\mathbf{x}_1, ..., \mathbf{x}_k$ be orthonormal, and assume that $\sum \alpha_j \mathbf{x}_j = \mathbf{o}$ for some scalars $\alpha_j$. Take the scalar product of both sides of this equation with each $\mathbf{x}_i$ successively, obtaining $\alpha_i = 0$ for $i = 1, ..., k$.

**1.14.** As in the text below (1.19), let $f(t) = |\mathbf{x} + t\mathbf{y}|^2$ for all real t, and assume that $|(\mathbf{x}, \mathbf{y})| = |\mathbf{x}|\,|\mathbf{y}|$. In this circumstance, one also has $f(t) = (|\mathbf{x}| \pm t|\mathbf{y}|)^2$. The result to be proved holds trivially if either $\mathbf{x}$ or $\mathbf{y}$ is null, so suppose neither is null. Then set $t = t_* \equiv \mp |\mathbf{x}|/|\mathbf{y}|$, so that $f(t_*) = 0$. It follows that $\mathbf{x} = \mp t_*\, \mathbf{y}$.

**1.15.** $|\mathbf{x} + \mathbf{y}|^2 = |\mathbf{x}|^2 + 2\,(\mathbf{x}, \mathbf{y}) + |\mathbf{y}|^2 \le |\mathbf{x}|^2 + 2\,|\mathbf{x}|\,|\mathbf{y}| + |\mathbf{y}|^2 = (|\mathbf{x}| + |\mathbf{y}|)^2$, so $|\mathbf{x} + \mathbf{y}| \le |\mathbf{x}| + |\mathbf{y}|$.

**1.16.** Let $\varphi(t) = \cos kt$, $\psi(t) = \sin kt$, with k a positive integer. Then $(\varphi, \psi) = \int_0^\pi \cos kt \sin kt\, dt = 0$, $|\varphi|^2 = \int_0^\pi \cos^2 kt\, dt = \int_0^\pi \sin^2 kt\, dt = |\psi|^2 = \pi/2 \neq 0$, so that the angle between $\varphi$ and $\psi$ is 90°.

**1.17.** (a) For i = 2, ..., k, prove that each of the new vectors $\mathbf{e}_i$ is orthogonal to all of its predecessors $\mathbf{e}_1, ..., \mathbf{e}_{i-1}$ by first showing through direct calculation that this proposition holds for i = 2, then establishing the general result by using mathematical induction. (b) In the space $\mathsf{A}_2$, the arrow $[(\mathbf{f}_2, \mathbf{e}_1)/(\mathbf{e}_1, \mathbf{e}_1)]\, \mathbf{e}_1$ is the orthogonal projection of the arrow $\mathbf{f}_2$ on the arrow $\mathbf{e}_1$, so that the arrow $\mathbf{e}_2$ delivered by the Gram-Schmidt process is perpendicular to $\mathbf{e}_1$.

**1.18.** Let $\mathbf{q}_0(t)$, $\mathbf{q}_1(t)$, $\mathbf{q}_2(t)$, and $\mathbf{q}_3(t)$ be the new polynomials produced from the **p**s by the Gram-Schmidt process using the scalar product specified in the problem. By direct calculation, one finds that $\mathbf{q}_0(t) = 1$, $\mathbf{q}_1(t) = t$, $\mathbf{q}_2(t) = t^2 - 1/3$ and $\mathbf{q}_3(t) = t^3 - 3t/5$. The first four Legendre polynomials are given by $P_0(t) = 1 = \mathbf{q}_0(t)$, $P_1(t) = t = \mathbf{q}_1(t)$, $P_2(t) = (3/2)t^2 - 1/2 = (3/2)\mathbf{q}_2(t)$, $P_3(t) = (5/2)t^3 - (3/2)t = (5/2)\mathbf{q}_3(t)$; for further discussion of these functions, see the book by R. Courant and D. Hilbert cited at the end of Chapter 2.

**1.19.** It is straightforward to verify that the proposed scalar product satisfies all of the rules. Note that, according to this scalar product, the "length" of a matrix is given by the square root of the sum of the squares of all of its elements.

**1.20.** (a) Let $\mathbf{x} = \{x_1, ..., x_n, ...\}$ and $\mathbf{y} = \{y_1, ..., y_n, ...\}$ be sequences in $l_2$. Using the natural operations to define $\alpha\mathbf{x}$ and $\mathbf{x} + \mathbf{y}$, where $\alpha$ is a scalar, we note that $\alpha\mathbf{x}$ is trivially in $l_2$ when $\mathbf{x}$ is. Next, an appeal to the Schwarz inequality shows that, for any integer $N > 1$,

$$\sum_{i=1}^{N}(x_i + y_i)^2 = \sum x_i^2 + \sum y_i^2 + 2\sum x_i y_i \leq \sum x_i^2$$

$$+ \sum y_i^2 + 2\,(\sum x_i^2)^{1/2}\,(\sum y_i^2)^{1/2},$$

so that $\sum_{i=1}^{\infty}(x_i + y_i)^2$ converges. Thus $\alpha\mathbf{x}$ and $\mathbf{x} + \mathbf{y}$ are in $l_2$ when $\mathbf{x}$ and $\mathbf{y}$ are in $l_2$, and $l_2$ is a vector space. (b) For any positive integer N, $\sum_{j=1}^{N}\alpha_j\, \mathbf{E}_j = \{\alpha_1, \alpha_2, ..., \alpha_N, 0, 0, ...\}$, which vanishes if and only if the $\alpha$s all vanish. Thus any finite set of the **E**s is a linearly independent set, and dim $l_2$ is infinity. (c) The only new issue here is whether $\sum_{i=1}^{\infty} x_i\, y_i$ converges; the argument using the Schwarz inequality in part (a) incidentally affirms this.

## Chapter 2

**2.1.** **A** is trivially linear. Let q be the null polynomial. Then $\alpha_N = \alpha_0 = \alpha_1 = \cdots \alpha_{N-1} = 0$, so p is necessarily the null polynomial as well. Thus the null space of **A** contains only the null polynomial, and **A** is nonsingular.

**2.2.** (a) $P_A(\lambda) = \lambda^2 - 2\lambda - 1$, so $\lambda_1 = 1 - 2^{1/2}$, $\lambda_2 = 1 + 2^{1/2}$. (b) $\underline{B}^{-1} = \begin{pmatrix} \cos\varphi & -\sin\varphi \\ \sin\varphi & \cos\varphi \end{pmatrix}$. (c) For any $\varphi$, $\underline{C} = \underline{B}^{-1}\underline{A}\,\underline{B} = \begin{pmatrix} 1 - \beta\sin 2\varphi & \beta\cos 2\varphi \\ \beta\cos 2\varphi & 1 + \beta\sin 2\varphi \end{pmatrix}$, where $\beta = 2^{1/2}$; choosing $\varphi = \pi/4$ yields the desired result.

**2.3.** (a) Since linear combinations of functions in $C_0^\infty$ (or $C_0^{\infty'}$) remain in $C_0^\infty$ (or $C_0^{\infty'}$), these two collections of functions are indeed linear vector spaces under the natural operations; $C_0^{\infty'}$ is a linear manifold in $C_0^\infty$. (b) The operation performed by **A** on input functions in $C_0^\infty$ in general yields output functions that are *not* in $C_0^\infty$, so **A** does not transform $C_0^\infty$ into itself. The same is true of **B**, since the derivative of a function in $C_0^\infty$ need not vanish at the origin. The operation performed by **C** does indeed yield outputs that vanish at the origin and are infinitely differentiable when the inputs have these properties. Also, $\mathbf{C}(\alpha f + \beta g) = \alpha\,\mathbf{C}f + \beta\,\mathbf{C}g$ for any real numbers $\alpha$, $\beta$ and any functions f, g in $C_0^\infty$. So **C** is a linear transformation of $C_0^\infty$ into itself. (c) When restricted to $C_0^{\infty'}$, **A** again fails to qualify as a transformation of the space into itself, but this is not true of **B**, which *is* a linear transformation of $C_0^{\infty'}$ into itself, as is **C**. (d) Considered as a linear transformation of $C_0^{\infty'}$ into itself, **C** is nonsingular, and $\mathbf{C}^{-1} = \mathbf{B}$. (e) This function is clearly infinitely differentiable for $t > 0$. Using the definition of derivative to check the differentiability of f at $t = 0$ yields the result that $f^{(n)}(0)$ exists and vanishes for every integer $n > 0$; this is because the derivative of order $n - 1$ of f(t) for $t > 0$ is of the form of a polynomial of degree n in 1/t multiplied by $\exp(-1/t)$. The latter dominates the former as t tends to zero from above.

**2.4.** One can show directly that the range of **AB** is the entire space, and the null space of **AB** contains only the null vector. Let $\mathbf{C} = \mathbf{B}^{-1}\mathbf{A}^{-1}$. Then $\mathbf{CAB} = \mathbf{B}^{-1}\mathbf{A}^{-1}\mathbf{A}\,\mathbf{B} = \mathbf{1}$.

**2.5.** For the matrix $\underline{A}$, one has $\det \underline{A} = 1 - \alpha$, so $\underline{A}$ is singular if and only if $\alpha = 1$. When $\alpha = 1$, the null space of $\underline{A}$ consists of all scalar multiples of $\begin{pmatrix} 1 \\ -1 \end{pmatrix}$, while the range of $\underline{A}$ is the set of all scalar multiples of $\begin{pmatrix} 1 \\ 1 \end{pmatrix}$.

**2.7.** Let $\underline{A}$ be an $n \times n$ upper triangular matrix with $a_{ij}$ in the $i^{th}$ row, $j^{th}$ column. It is immediate that $\det(\underline{A} - \lambda\,\underline{1}) = (a_{11} - \lambda)(a_{22} - \lambda)\cdots(a_{nn} - \lambda)$, so that the eigenvalues are given by $\lambda_1 = a_{11}$, $\lambda_2 = a_{22}$, $\cdots$, $\lambda_n = a_{nn}$. Clearly $\det \underline{A} = a_{11} \cdots a_{nn}$, so $\underline{A}$ is nonsingular if and only if all of its diagonal elements are nonzero. When $n = 3$ and $\underline{A}$ is nonsingular, its inverse is the upper triangular matrix given by

$$\underline{A}^{-1} = \begin{pmatrix} 1/a_{11} & -a_{12}/(a_{11}a_{22}) & (a_{12}a_{23} - a_{13}a_{22})/(a_{11}a_{22}a_{33}) \\ 0 & 1/a_{22} & -a_{23}/(a_{22}a_{33}) \\ 0 & 0 & 1/a_{33} \end{pmatrix}.$$

**2.8.** If p(t) is continuous for $t \geq 0$ and satisfies (2.20), then it is also differentiable for $t > 0$, and satisfies $\lambda\, t\, p'(t) + (\lambda - 1)p(t) = 0$. The case $\lambda = 0$ is clearly of no interest; if $\lambda \neq 0$, then the differential equation just cited implies that $p(t) = C\, t^{-1+1/\lambda}$ for some constant C. It follows that $\lambda$ must satisfy $0 < \lambda \leq 1$. Conversely, if $\lambda$ is in this range, one can verify directly that p(t) as given above is continuous for $t \geq 0$ and satisfies (2.20). Note that, in the discussion given in Example 2.9 where p(t) is assumed to be a polynomial of degree not exceeding n, the eigenvalues all lie in the interval (1/n, 1]. As n tends to infinity, this interval tends to (0, 1], and the number of eigenvalues in Example 2.9 tends to infinity.

**2.9.** Since the space is finite dimensional, it is sufficient to show that the null space of **R** contains only the null vector. Suppose $\mathbf{Rx} = \mathbf{o}$. Since the **e**s form a basis, we can write $\mathbf{x} = \alpha_j\, \mathbf{e}_j$, so that $\mathbf{Rx} = \alpha_j\, \mathbf{Re}_j = \alpha_j\, \mathbf{f}_j = \mathbf{o}$; remember the summation convention. But since the **f**s form a basis, they are linearly independent, so the $\alpha$s must all vanish. Thus $\mathbf{x} = \mathbf{o}$.

**2.10.** Since the cases $k = 0$ and $m = 0$ are trivial, assume $k > 0$, $m > 0$. Let $\mathbf{e}_1, ..., \mathbf{e}_m$ be a basis for $N(\mathbf{A})$. Since $N(\mathbf{A}) \neq \mathsf{R}$, there is a vector in $\mathsf{R}$ that is not in $N(\mathbf{A})$; call this vector $\mathbf{e}_{m+1}$. Since $N(\mathbf{A})$ is a linear manifold in $\mathsf{R}$, $\mathbf{e}_1, ..., \mathbf{e}_m, \mathbf{e}_{m+1}$ must be linearly independent. Put $\mathsf{M}_1 = \text{span}\,\{\mathbf{e}_1, ..., \mathbf{e}_{m+1}\}$. If $m + 1 \neq n$, there is a vector in $\mathsf{R}$ that is not in $\mathsf{M}_1$, say $\mathbf{e}_{m+2}$. The vectors $\mathbf{e}_1, ..., \mathbf{e}_{m+2}$ must be independent; let $\mathsf{M}_2$ be their span. Continue in this way, ultimately generating a set $\mathbf{e}_1, \mathbf{e}_2, ..., \mathbf{e}_m, \mathbf{e}_{m+1}, ..., \mathbf{e}_n$ that extends the basis for $N(\mathbf{A})$ to a basis for *all* of $\mathsf{R}$. Now let **z** be in $R(\mathbf{A})$, so that there is a vector **w** in $\mathsf{R}$ such that $\mathbf{Aw} = \mathbf{z}$; write $\mathbf{w} = \sum_{i=1}^{n} \alpha_i\, \mathbf{e}_i$ for suitable scalars $\alpha$. Since the first m **e**s are in $N(\mathbf{A})$, one has $\mathbf{z} = \sum_{i=m+1}^{n} \alpha_i\, \mathbf{f}_i$, where $\mathbf{f}_i = \mathbf{Ae}_i$, $i = m + 1, ..., n$. Thus every vector in $R(\mathbf{A})$ is a linear combination of the **f**s. Suppose the vectors $\mathbf{f}_{m+1}, ..., \mathbf{f}_n$ were linearly dependent; then there would be scalars $\beta_{m+1}, ..., \beta_n$ such that $\sum_{i=m+1}^{n} \beta_i\, \mathbf{f}_i = \mathbf{o}$, or $\mathbf{A}(\sum_{i=m+1}^{n} \beta_i\, \mathbf{e}_i) = \mathbf{o}$. But since $\mathbf{e}_{m+1}, ..., \mathbf{e}_n$ are linearly independent and $\sum \beta_i\, \mathbf{e}_i$ is not in $N(\mathbf{A})$ unless it is null, all the $\beta$s must vanish. Therefore the **f**s, of which there are $n - m$, form a basis for $R(\mathbf{A})$, and $k = \dim R(\mathbf{A}) = n - m$.

**2.11.** Let **e** be an eigenvector of **A** corresponding to the eigenvalue $\lambda$, so $\mathbf{Ae} = \lambda\mathbf{e}$, and set $\mathbf{f} = \mathbf{R}^{-1}\mathbf{e}$. Then $\mathbf{Bf} = \mathbf{R}^{-1}\mathbf{ARR}^{-1}\mathbf{e} = \mathbf{R}^{-1}\mathbf{Ae} = \lambda\mathbf{R}^{-1}\mathbf{e} = \lambda\mathbf{f}$, so **f** is an eigenvector of **B** corresponding to the eigenvalue $\lambda$. (Is **f** really non-null?) A similar argument shows that every eigenvalue of **B** is an eigenvalue of **A**, with corresponding eigenvectors related by $\mathbf{f} = \mathbf{R}^{-1}\mathbf{e}$ (*). Hence the eigenvalues of **A** and **B** coincide, and corresponding eigenvectors are related by (*).

**2.12.** (a) (***) follows immediately from the facts that Tr $\underline{B}^{(1)}$ = Tr $\underline{B}^{(2)}$, det $\underline{B}^{(1)}$ = det $\underline{B}^{(2)}$. (b) (***) implies that $\nu^2 = \alpha\delta - \beta\gamma - (\alpha + \delta)^2/4 = -(1/4)(\alpha - \delta)^2 - \beta\gamma$, so if $\nu \neq 0$, one has $(\alpha - \delta)^2 + 4\beta\gamma < 0$. (c) If **B** has no one-dimensional invariant linear manifold, then the discriminant of the characteristic polynomial $P_{\mathbf{B}}(\lambda)$ must be negative; this is precisely the inequality established in part (b). (d) Clearly, (*) is equivalent to a homogeneous linear system of four equations for the four unknown elements of the matrix $\underline{R}$. (e) With the help of (***), one can show that this system has a vanishing determinant, and that one of its solutions leads to the choice

$$\underline{R} = \begin{pmatrix} \delta - \alpha & [-(\delta - \alpha)^2 - 4\beta\gamma]^{1/2} \\ -2\gamma & 0 \end{pmatrix}$$

for the matrix $\underline{R}$, which is non-singular because of the inequality established in part (b). (f) By referring **B** to the basis **u**, **v**, one immediately finds (****).

**2.14.** Suppose there is a non-null vector **z** in M such that $\mathbf{Az} = \lambda\mathbf{z}$ for some real $\lambda$. Since **x**, **y** form a basis for M, one can write $\mathbf{z} = \alpha\mathbf{x} + \beta\mathbf{y}$ for some scalars $\alpha$, $\beta$. Using (2.44), one then finds that $\alpha$, $\beta$ satisfy $(\mu - \lambda)\alpha + \nu\beta = 0$, $-\nu\,\alpha + (\mu - \lambda)\beta = 0$. Because $\nu \neq 0$, the determinant of this system cannot vanish for any real choice of $\lambda$. Thus $\alpha = \beta = 0$, from which it follows that $\mathbf{z} = \mathbf{o}$, which is a contradiction.

**2.15.** In terms of components in any basis, Tr (**AB**) = $a_{ij}\, b_{ji}$ and Tr (**BA**) = $b_{ij}a_{ji} = b_{ji}\, a_{ij} = a_{ij}\, b_{ji}$ = Tr (**AB**).

**2.16.** If there is a basis **f** in which the matrix $\underline{A}^f$ of **A** is diagonal, say $\underline{A}^f = \begin{pmatrix} \lambda_1 & 0 \\ 0 & \lambda_2 \end{pmatrix}$, then $\lambda_1$ and $\lambda_2$ are the eigenvalues of **A**. But there is a basis **e** in which $\underline{A}^e = \begin{pmatrix} 1 & 1 \\ 0 & 1 \end{pmatrix}$, so that the eigenvalues of **A** are in fact $\lambda_1 = \lambda_2 = 1$. Thus $\underline{A}^f = \underline{1}$, so **A** must be the identity, which is a contradiction.

**2.18.** (a) Since the vectors $\mathbf{e}_1, ..., \mathbf{e}_n$ of the basis **e** are necessarily eigenvectors of **A**, one has $\mathbf{A}^k\mathbf{e}_i = \lambda_i\, \mathbf{A}^{k-1}\mathbf{e}_i = \cdots = \lambda_i^{\,k}\, \mathbf{e}_i$ (no sum on i), so that the matrix of components of $\mathbf{A}^k$ is also diagonal in the basis **e**, with diagonal elements $\lambda_1^{\,k}, \lambda_2^{\,k}, ..., \lambda_n^k$. (b) If $\mathbf{B} = p(\mathbf{A})$ is any polynomial in **A**, then the matrix of components of **B** in **e** is also diagonal, with diagonal elements $p(\lambda_1), ..., p(\lambda_n)$. (c) Choose $p(\mathbf{A}) = P_{\mathbf{A}}(\mathbf{A})$, where $P_A(\lambda)$ is the characteristic polynomial of $\mathbf{A}$. Then for each i = 1, ..., n, one has $p(\mathbf{A})\ \mathbf{e}_i = p(\lambda_i)\ \mathbf{e}_i = \mathbf{o}$ (no sum on i), so that $p(\mathbf{A})$ is the null tensor. Thus $P_A(\mathbf{A}) = \mathbf{O}$.

**2.19.** Write the characteristic polynomial of $\mathbf{A}$ as $P_A(\lambda) = \sum_{k=0}^{n} p_k\ \lambda^k$, and note that $p_0 = \det \mathbf{A}$. By the Cayley-Hamilton theorem (see Problem 2.18), one has $p_0\mathbf{1} = -\sum_{k=1}^{n} p_k\ \mathbf{A}^k$. If $\mathbf{A}$ is nonsingular, then $p_0 \neq 0$, so that $\mathbf{A}^{-1} = -(1/p_0) \sum_{k=1}^{n} p_k\ \mathbf{A}^{k-1}$.

**2.20.** In a two-dimensional space, the characteristic polynomial of any tensor $\mathbf{C}$ is given by $P_C(\lambda) = \lambda^2 - (\mathrm{Tr}\ \mathbf{C})\ \lambda + \det \mathbf{C}$. Therefore by the Cayley-Hamilton theorem, one has

$$\mathbf{C}^2 - (\mathrm{Tr}\ \mathbf{C})\ \mathbf{C} + (\det \mathbf{C})\ \mathbf{1} = 0.$$

But if $\mathbf{C} = \mathbf{AB} - \mathbf{BA}$, then $\mathrm{Tr}\ \mathbf{C} = 0$ (see Problem 2.15), so that $\mathbf{C}^2 = -(\det \mathbf{C})\ \mathbf{1}$.

## Chapter 3

**3.1.** (a) The result follows from the linearity of the scalar product in its arguments. (b) The average value of a continuous function $\varphi$ on $[0, \pi]$ is $(1/\pi) \int_0^{\pi} \varphi(t)\ dt = (h, \varphi)$, where h is the (constant) function $h(t) = 1/\pi$, $0 \le t \le \pi$. (c) By the linearity of $\psi$, one has $\psi(\mathbf{x}) = \psi(\xi_i \mathbf{e}_i) = \xi_i \psi(\mathbf{e}_i) = \xi_i h_i = (\mathbf{h}, \mathbf{x})$. If $\psi(\mathbf{x}) = (\mathbf{h}_1, \mathbf{x}) = (\mathbf{h}_2, \mathbf{x})$, then $\mathbf{h}_1 - \mathbf{h}_2$ is orthogonal to every vector in R, and therefore vanishes. Thus $\mathbf{h}$ is unique.

**3.2.** Clearly, $(\mathbf{x}, \mathbf{Cy}) = (\mathbf{x}, \mathbf{A}(\mathbf{By})) = (\mathbf{A}^T\mathbf{x}, \mathbf{By}) = (\mathbf{B}^T\mathbf{A}^T\mathbf{x}, \mathbf{y})$, so that $\mathbf{C}$ indeed has a transpose, and it is given by $\mathbf{C}^T = \mathbf{B}^T\mathbf{A}^T$.

**3.3.** Let $\mathbf{C} = \mathbf{a}\otimes\mathbf{b}$; then $(\mathbf{x}, \mathbf{Cy}) = (\mathbf{x}, (\mathbf{b},\mathbf{y})\mathbf{a}) = (\mathbf{b},\mathbf{y})(\mathbf{x},\mathbf{a}) = (\mathbf{y}, (\mathbf{a},\mathbf{x})\mathbf{b}) = (\mathbf{Dx}, \mathbf{y})$, where $\mathbf{D} = \mathbf{b}\otimes\mathbf{a}$. It follows that $\mathbf{D} = \mathbf{C}^T$, or $(\mathbf{a}\otimes\mathbf{b})^T = \mathbf{b}\otimes\mathbf{a}$. Thus $\mathbf{C}$ is symmetric if and only if $\mathbf{a}\otimes\mathbf{b} = \mathbf{b}\otimes\mathbf{a}$; it is easily shown that this holds if and only if $\mathbf{a}$ and $\mathbf{b}$ are linearly dependent, from which the desired result follows.

**3.4.** For any $\mathbf{x}$ in R, $\mathbf{Cx} = \mathbf{A}(\mathbf{b}\otimes\mathbf{b})\mathbf{x} = (\mathbf{b}, \mathbf{x})\mathbf{Ab} = (\mathbf{b}, \mathbf{x})\ (\mathbf{a}\otimes\mathbf{a})\mathbf{b} = (\mathbf{b}, \mathbf{x})\ (\mathbf{a}, \mathbf{b})\mathbf{a} = \mathbf{o}$, so $\mathbf{C} = \mathbf{O}$. Also, $\mathbf{Dx} = \mathbf{A}^2\mathbf{x} = \mathbf{A}(\mathbf{a}\otimes\mathbf{a})\mathbf{x} = (\mathbf{a}, \mathbf{x})\ \mathbf{Aa} = (\mathbf{a}, \mathbf{x})(\mathbf{a}\otimes\mathbf{a})\mathbf{a} = (\mathbf{a}, \mathbf{x})(\mathbf{a}, \mathbf{a})\mathbf{a} = (\mathbf{a}, \mathbf{x})\mathbf{a} = (\mathbf{a}\otimes\mathbf{a})\mathbf{x} = \mathbf{Ax}$ for every $\mathbf{x}$ in R, so $\mathbf{A}^2 = \mathbf{A}$.

**3.5.** If $(\mathbf{a}\otimes\mathbf{b})\mathbf{x} = \lambda\mathbf{x}$, then $(\mathbf{b}, \mathbf{x})\ \mathbf{a} = \lambda\ \mathbf{x}$ (*). Suppose first that $\lambda = 0$. Then $(\mathbf{b}, \mathbf{x}) = 0$, so $\mathbf{x}$ is orthogonal to $\mathbf{b}$. Conversely, any non-null vector $\mathbf{x}$ that is orthogonal to $\mathbf{b}$ is an eigenvector of $\mathbf{a}\otimes\mathbf{b}$ corresponding to the eigenvalue $\lambda = 0$. Now suppose $\lambda \neq 0$. Then (*) above shows that $\mathbf{x}$ is a scalar multiple of **a**, say $\mathbf{x} = \alpha\,\mathbf{a}$, $\alpha \neq 0$; (*) yields $\lambda = (\mathbf{a}, \mathbf{b})$. Thus there are at most two distinct eigenvalues; all non-null vectors orthogonal to **b** are eigenvectors corresponding to the eigenvalue $\lambda = 0$, and all non-null scalar multiples of the vector **a** are eigenvectors corresponding to $\lambda = (\mathbf{a}, \mathbf{b})$. What happens if **a** is orthogonal to **b**?

**3.6.** (a) Let **x** be in the null space of $\mathbf{a}\otimes\mathbf{b}$, so that $(\mathbf{b}, \mathbf{x})\,\mathbf{a} = \mathbf{o}$. If $\mathbf{a} = \mathbf{o}$, it is clear that *all* vectors **x** are in $N(\mathbf{A})$, so dim $N(\mathbf{A}) = n$, and $r(\mathbf{A}) = 0$. If $\mathbf{a} \neq \mathbf{o}$, then $N(\mathbf{A})$ consists of the set of all vectors orthogonal to **b**, so dim $N(\mathbf{A}) = n - 1$, and $r(\mathbf{A}) = 1$. (b) Suppose $r(\mathbf{A}) = 1$, so that dim $N(\mathbf{A}) = n - 1$. . Let $\mathbf{x}_1, \ldots, \mathbf{x}_{n-1}$ be a basis for $N(\mathbf{A})$. Let **b** be a unit vector orthogonal to all $n - 1$ of these **x**s, and set $\mathbf{a} = \mathbf{Ab}$. Using the fact that the $\mathbf{x}_i$s and **b** form a basis for $\mathsf{R}$, one readily shows that $\mathbf{Ax} = (\mathbf{x}, \mathbf{b})\,\mathbf{Ab} = (\mathbf{x}, \mathbf{b})\,\mathbf{a} = (\mathbf{a}\otimes\mathbf{b})\mathbf{x}$ for all **x** in $\mathsf{R}$, so that $\mathbf{A} = \mathbf{a}\otimes\mathbf{b}$. Since **b** is not in $N(\mathbf{A})$, necessarily $\mathbf{a} \neq \mathbf{o}$.

**3.7.** Let **x** be any non-null vector orthogonal to **b**. It is given that $\mathbf{A} = \mathbf{1} + \mathbf{a}\otimes\mathbf{b}$ for some vectors **a**, **b**, so $\mathbf{Ax} = \mathbf{x} + (\mathbf{a}\otimes\mathbf{b})\mathbf{x} = \mathbf{x} + (\mathbf{b}, \mathbf{x})\mathbf{a} = \mathbf{x}$; thus **x** is an eigenvector of **A** corresponding to the eigenvalue one.

**3.8.** $\mathbf{Ae}_1 = a^{\mathbf{e}}_{11}\,\mathbf{e}_1 + a^{\mathbf{e}}_{21}\,\mathbf{e}_2$; also, $\mathbf{Ae}_1 = \begin{pmatrix} 0 & 1 \\ 1 & 0 \end{pmatrix}\begin{pmatrix} 1 \\ 0 \end{pmatrix} = \begin{pmatrix} 0 \\ 1 \end{pmatrix} = \mathbf{e}_2$, so $a^{\mathbf{e}}{}_{11} = 0$, $a^{\mathbf{e}}_{21} = 1$. Similarly, one finds $a^{\mathbf{e}}_{12} = 1$, $a^{\mathbf{e}}_{22} = 1$; thus $\underline{A}^{\mathbf{e}} = \begin{pmatrix} 0 & 1 \\ 1 & 0 \end{pmatrix}$. Also, $\mathbf{Af}_1 = \begin{pmatrix} 0 & 1 \\ 1 & 0 \end{pmatrix}\begin{pmatrix} 1 \\ 0 \end{pmatrix} = \mathbf{e}_2 = -\mathbf{f}_1 + \mathbf{f}_2$, $\mathbf{Af}_2 = \mathbf{f}_2$, so $\underline{A}^{\mathbf{f}}\begin{pmatrix} -1 & 1 \\ 1 & 0 \end{pmatrix}$. **A** is symmetric, because $(\mathbf{Ax}, \mathbf{y}) = (\mathbf{x}, \mathbf{Ay})$, as is easily verified. Since **e** is an orthonormal basis in the natural scalar product, the matrix $\underline{A}^{\mathbf{e}}$ is a symmetric one ($a^{\mathbf{e}}_{12} = a^{\mathbf{e}}_{21}$). On the other hand, **f** is *not* an orthonormal basis, and indeed $\underline{A}^{\mathbf{f}}$ is not a symmetric matrix.

**3.9. Let** $e_1, \ldots, \mathbf{e}_n$ be a principal basis for **A**, and represent **x** as $\mathbf{x} = \sum \xi_i \mathbf{e}_i$. Then $\mathbf{Ax} = \sum \xi_i \mathbf{Ae}_i = \sum \lambda_i \xi_i\, \mathbf{e}_i$, where the $\lambda_i$'s are the eigenvalues of A, and therefore $f(\mathbf{x}) = \sum \lambda_i\, \xi_i^2$. If **A** is symmetric, then all the $\lambda_i$s are positive, so $f(\mathbf{x}) > 0$ for all non-null **x**. Conversely, if f has this property, then no $\lambda_i$ can fail to be positive. Indeed suppose $\lambda_k \leq 0$; then $f(\mathbf{e}_k) = \lambda_k \leq 0$ (no sum), contradicting the positivity of $f(\mathbf{x})$ for $\mathbf{x} \neq \mathbf{o}$.

**3.10.** If **A** is skew-symmetric and $\mathbf{Ax} = \lambda\mathbf{x}$, then $\lambda(\mathbf{x}, \mathbf{x}) = (\mathbf{x}, \mathbf{Ax}) = (\mathbf{A}^T\mathbf{x}, \mathbf{x}) = \; = -(\mathbf{Ax}, \mathbf{x}) = -\lambda(\mathbf{x}, \mathbf{x})$, so $\lambda\,(\mathbf{x}, \mathbf{x}) = 0$. If $\lambda = 0$ and $\mathbf{x} \neq \mathbf{o}$, then **A** is singular, since $\mathbf{Ax} = \mathbf{o}$. If $\lambda \neq 0$, then necessarily $\mathbf{x} = \mathbf{o}$, so **A** can have no nonzero eigenvalue. In the real Euclidean space $\mathsf{R}_2$ of 2-columns, the transformation $\begin{pmatrix} 0 & 1 \\ -1 & 0 \end{pmatrix}$ is skew symmetric, non-singular and has no eigenvalue; indeed, *every* non-null skew-symmetric transformation on $\mathsf{R}_2$ is non-singular and has no eigenvalue. In $\mathsf{R}_3$, the transformation $\begin{pmatrix} 0 & 1 & 0 \\ -1 & 0 & 0 \\ 0 & 0 & 0 \end{pmatrix}$ is skew-symmetric, singular, and therefore has eigenvalue zero.

**3.12.** (a) If $\mathbf{x}$ is in M, then $\mathbf{x} = \sum_{j=1}^{k} \xi_j \mathbf{e}_j$ for suitable scalars $\xi_j$, and $\mathbf{Px}$, being a linear combination of the $\mathbf{e}_j$s, is also in M. (b) For any $\mathbf{x}$, $\mathbf{y}$ in R, one has $(\mathbf{Px}, \mathbf{y}) = \Sigma\ (\mathbf{x}, \mathbf{e}_j)(\mathbf{y}, \mathbf{e}_j) = (\mathbf{x}, \mathbf{Py})$, so $\mathbf{P}^T = \mathbf{P}$. Also, obviously $\mathbf{P}^2 = \mathbf{P}$. It then follows that $(\mathbf{Px}, \mathbf{x} - \mathbf{Px}) = (\mathbf{Px}, \mathbf{x}) - (\mathbf{Px}, \mathbf{Px})$. But $(\mathbf{Px}, \mathbf{Px}) = (\mathbf{x}, \mathbf{P}^T\mathbf{Px}) = (\mathbf{x}, \mathbf{P}^2\mathbf{x}) = (\mathbf{x}, \mathbf{Px})$, so $(\mathbf{Px}, \mathbf{x} - \mathbf{Px}) = 0$. (c) For any $\mathbf{x}$ in R, $(\mathbf{n}\otimes\mathbf{n})\ \mathbf{x} = (\mathbf{n}, \mathbf{x})\ \mathbf{n}$, which is clearly a projection; M is the set of all scalar multiples of $\mathbf{n}$. (d) Using the result of part (b), $0 \le |\mathbf{x} - \mathbf{Px}|^2 = (\mathbf{x} - \mathbf{Px},\ \mathbf{x} - \mathbf{Px})^2 = (\mathbf{x},\ \mathbf{x}) - (\mathbf{x},\ \mathbf{Px}) = (\mathbf{x},\ \mathbf{x}) - (\mathbf{x},\ \mathbf{P}^2\mathbf{x}) = (\mathbf{x}, \mathbf{x}) - (\mathbf{Px}, \mathbf{Px}) = |\mathbf{x}|^2 - |\mathbf{Px}|^2 = |\mathbf{x}|^2 - \sum_{j=1}^{k} (\mathbf{x}, \mathbf{e}_j)^2$, establishing the result. (e) By part (d), $\sum_{j=1}^{k}(\mathbf{x}, \mathbf{e}_j)^2 \le |\mathbf{x}|^2$ for every positive integer k. Since the right side of the inequality is independent of k and the left side does not decrease with increasing k, the series $\sum_{j=1}^{\infty}(\mathbf{x}, \mathbf{e}_j)^2$ converges.

**3.13.** (a) If $\mathbf{Ax} = \mathbf{b}$, then $\sum \xi_i \mathbf{Ae}_i = \sum \xi_i \lambda_i \mathbf{e}_i = \sum b_i \mathbf{e}_i$, so $\lambda_i \xi_i = b_i$ (no sum), $i = 1, ..., n$ (*). Conversely, if (*) holds, so does $\mathbf{Ax} = \mathbf{b}$. (b) If all the $\lambda_i$s are nonzero, (*) gives $\xi_i = b_i/\lambda_i$ (no sum), so that $\mathbf{x} = \sum (b_i/\lambda_i)\mathbf{e}_i$ is a solution of $\mathbf{Ax} = \mathbf{b}$. Since $\mathbf{A}$ is nonsingular if and only if all of its eigenvalues are nonzero, the solution is unique under this assumption. (c) In this case, (*) cannot hold unless $b_1 = b_2 = \cdots = b_k = 0$, or $(\mathbf{b}, \mathbf{e}_i) = 0$, $i = 1, ..., k$ (**). Conversely if (**) holds, then (*) yields $\xi_i = b_i/\lambda_i$ for $i = k + 1, k + 2, ..., n$, but leaves $\xi_1 = \xi_2 = \cdots = \xi_k$ arbitrary. Thus in this case, there is a solution of $\mathbf{Ax} = \mathbf{b}$ if and only if $(\mathbf{b}, \mathbf{e}_i) = 0$ for $i = 1, ..., k$, and the solution is determined only to within an arbitrary linear combination of the first k $\mathbf{e}_i$s. Parts (d) and (e) are restatements of the results established in parts (a)–(c).

**3.15.** In terms of the components of $\mathbf{x}$ in the natural basis, one finds that the quadratic form is given by $f(\mathbf{x}) = x_1^2 - 2x_1x_2 + 2x_2^2 - 2x_2x_3 + 2x_3^2 - 2x_3x_4 + x_4^2$. Choose $x_1 = 1$, $x_2 = x_3 = x_4 = 0$. By Rayleigh's principle as established in Problem 3.14, one has $\lambda_1 \le 1$. A better bound is obtained by retaining $x_1 = 1$, but choosing $x_2$, $x_3$, $x_4$ so as to minimize $f(\mathbf{x})$ under the constraint that $x_1 = 1$. This leads to $\lambda_1 \le 0$. The exact eigenvalues are 0, $2 - 2^{1/2}$, 2, and $2 + 2^{1/2}$.

**3.16.** If the largest eigenvalue $\lambda$ of $\mathbf{A}$ is such that $|\lambda| \ge 1$, then the corresponding normalized eigenvector $\mathbf{e}$ is such that $|\mathbf{Ae}| = |\lambda| \ge 1$, so that $\mathbf{A}$ cannot be a contraction. It follows that $\lambda < 1$ is necessary for $\mathbf{A}$ to be a contraction. If the largest eigenvalue $\lambda$ of $\mathbf{A}$ is less than one, then

$(\mathbf{Ax}, \mathbf{x}) \leq \lambda\, |\mathbf{x}|^2$ for every $\mathbf{x}$ in R; choose $\alpha = \lambda$. It follows that $\mathbf{A}$ is a contraction.

**3.17.** Since $\mathbf{AB} = \mathbf{BA}$, $\mathbf{A}$ and $\mathbf{B}$ do indeed have a common principal basis.

**3.18.** For any $\mathbf{x}$ in R, $|\mathbf{Qx}|^2 - |\mathbf{x}|^2 = (\mathbf{Qx}, \mathbf{Qx}) - (\mathbf{x}, \mathbf{x}) = (\mathbf{x}, (\mathbf{Q}^T\mathbf{Q} - \mathbf{1})\mathbf{x})$. It follows that $\mathbf{Q}$ preserves the length of every vector in R if and only if $\mathbf{Q}^T\mathbf{Q} = \mathbf{1}$.

**3.19.** By (3.30), there are linearly independent vectors $\mathbf{f}$, $\mathbf{g}$ in M such that $\mathbf{Qf} = \cos\varphi\,\mathbf{f} - \sin\varphi\,\mathbf{g}$, and $\mathbf{Qg} = \sin\varphi\,\mathbf{f} + \cos\varphi\,\mathbf{g}$ for some real angle $\varphi$. Operating on each of these equations with $\mathbf{Q}^T$ yields $\mathbf{f} = \cos\varphi\,\mathbf{Q}^T\mathbf{f} - \sin\varphi\,\mathbf{Q}^T\mathbf{g}$, $\mathbf{g} = \sin\varphi\,\mathbf{Q}^T\mathbf{f} + \cos\varphi\,\mathbf{Q}^T\mathbf{g}$. Solving these equations for $\mathbf{Q}^T\mathbf{f}$ and $\mathbf{Q}^T\mathbf{g}$ shows that M is also a two-dimensional invariant linear manifold for $\mathbf{Q}^T$ corresponding to the zero $\exp(-i\varphi)$ of the characteristic polynomial of $\mathbf{Q}^T$.

**3.20.** By the orthogonality condition satisfied by the components of an orthogonal tensor in an orthonormal basis, one has $q_{i1}^2 + q_{i2}^2 + \cdots + q_{in}^2 = 1$ for $i = 1, \ldots, n$. The desired result follows immediately.

**3.21.** According to the discussion leading to Proposition 3.5, the invariant linear manifold K corresponding to the eigenvalue $\lambda = 1$ (if it exists) of an orthogonal tensor $\mathbf{Q}$ has dimension k. From the representation (3.35) for the matrix $\underline{Q}$, it is clear that $\lambda = 1$ is a zero of multiplicity k of the characteristic polynomial of $\mathbf{Q}$. Thus the geometric and algebraic multiplicities of the eigenvalue $\lambda = 1$ coincide. The same is true of the eigenvalue $\lambda = -1$.

**3.22.** Let $\lambda$ be an eigenvalue of $\mathbf{V}$ with corresponding eigenvector $\mathbf{e}$: $\mathbf{Ve} = \lambda\mathbf{e}$. Then if $\mathbf{R}$ is an orthogonal tensor, $\mathbf{R}^T\mathbf{VRR}^T\mathbf{e} = \lambda\,\mathbf{R}^T\mathbf{e}$, so $\mathbf{R}^T\mathbf{e}$ is an eigenvector of $\mathbf{R}^T\mathbf{VR}$ with the same eigenvalue $\lambda$. Choose $\mathbf{R}$ to be the orthogonal tensor for which $\mathbf{U} = \mathbf{R}^T\mathbf{VR}$ in the polar decomposition to conclude that the stretch factors $\mathbf{V}$ and $\mathbf{U}$ have the same eigenvalues.

**3.23.** Let $\mathbf{f}_1$, $\mathbf{f}_2$ be the natural basis in $R_2$. $(\underline{U}^f)^2 = (\underline{A}^f)^T\underline{A}^f = \begin{pmatrix} 5 & -2 \\ -2 & 8 \end{pmatrix}$. The eigenvalues $\lambda$ and eigenvectors $\mathbf{e}$ of $\mathbf{U}^2$ are found by using the above matrix to be $\lambda_1 = 4$, $\lambda_2 = 9$, $\mathbf{e}_1 = 5^{-1/2}\begin{pmatrix} 2 \\ 1 \end{pmatrix}$, $\mathbf{e}_2 = 5^{-1/2}\begin{pmatrix} 1 \\ -2 \end{pmatrix}$. Determine $\mathbf{U}$ by setting $\underline{U}^e = \begin{pmatrix} 2 & 0 \\ 0 & 3 \end{pmatrix}$. Let $\underline{Q}^e = 5^{-1/2}\begin{pmatrix} 2 & 1 \\ 1 & -2 \end{pmatrix}$, so that $\underline{U}^f = (\underline{Q}^e)^T\,\underline{U}^e\underline{Q}^e = (1/5)\begin{pmatrix} 11 & -2 \\ -2 & 14 \end{pmatrix}$. Next, find $(\underline{U}^f)^{-1} = (1/15)\begin{pmatrix} 7 & 1 \\ 1 & 11 \end{pmatrix}$. Use the right polar decomposition $\mathbf{A} = \mathbf{RU}$ to get $\underline{R}^f = \underline{A}^f(\underline{U}^f)^{-1} = (1/5)\begin{pmatrix} 3 & 4 \\ -4 & 3 \end{pmatrix}$. Thus the right polar decomposition of $\mathbf{A}$ is $\begin{pmatrix} 1 & 2 \\ -2 & 2 \end{pmatrix} = \begin{pmatrix} 3/5 & 4/5 \\ -4/5 & 3/5 \end{pmatrix}\begin{pmatrix} 11/5 & -2/5 \\ -2/5 & 14/5 \end{pmatrix}$.

Since the factor **V** in the left polar decomposition satisfies $\mathbf{V} = \mathbf{R}\mathbf{U}\mathbf{R}^T$, one readily finds that $\underline{V}^f = (1/5)\begin{pmatrix} 11 & 2 \\ 2 & 14 \end{pmatrix}$.

**3.24.** Since **A** is symmetric and positive definite, there is a basis in which its matrix is diagonal, with the (positive) eigenvalues of **A** as the diagonal elements. Since **A** is orthogonal, each column has unit "length," so every eigenvalue is either 1 or $-1$. But for **A** to be positive definite, all of its eigenvalues must be positive, so **A** must be the identity tensor.

**3.26.** Suppose $\mathbf{1} + \mathbf{A}$ is singular. Then there is a non-null vector **x** such that $\mathbf{A}\mathbf{x} = -\mathbf{x}$, so that $\lambda = -1$ is an eigenvalue of **A**. But it was shown in Problem 3.10 that the only possible eigenvalue of a skew-symmetric tensor is zero. Hence $\mathbf{1} + \mathbf{A}$ is non-singular; similarly $\mathbf{1} - \mathbf{A}$ is also non-singular. Set $\mathbf{Q} = (\mathbf{1} - \mathbf{A})(\mathbf{1} + \mathbf{A})^{-1}$. Then $\mathbf{Q}^T = (\mathbf{1} - \mathbf{A})^{-1}(\mathbf{1} + \mathbf{A})$, and therefore one has $\mathbf{Q}^T\mathbf{Q} = (\mathbf{1} - \mathbf{A})^{-1}(\mathbf{1} + \mathbf{A})(\mathbf{1} - \mathbf{A})(\mathbf{1} + \mathbf{A})^{-1} = (\mathbf{1} - \mathbf{A})^{-1}(\mathbf{1} - \mathbf{A}^2)(\mathbf{1} + \mathbf{A})^{-1} = (\mathbf{1} - \mathbf{A})^{-1}(\mathbf{1} - \mathbf{A})(\mathbf{1} + \mathbf{A})(\mathbf{1} + \mathbf{A})^{-1} = \mathbf{1}$.

## Chapter 4

**4.1.** Let $\mathbf{L} = \mathbf{a}\otimes\mathbf{b}$, $\mathbf{M} = \mathbf{c}\otimes\mathbf{d}$, where **a**, **b**, **c**, and **d** are vectors. Then in terms of components in an orthonormal basis $\mathbf{e}_1, \ldots, \mathbf{e}_n$, $\langle \mathbf{L}, \mathbf{M} \rangle = \text{Tr}\,(\mathbf{L}\mathbf{M}^T) = a_p b_q d_q c_p = (\mathbf{a}, \mathbf{c})(\mathbf{b}, \mathbf{d})$. If $\mathbf{a} = \mathbf{e}_i$, $\mathbf{b} = \mathbf{e}_j$, $\mathbf{c} = \mathbf{e}_k$, $\mathbf{d} = \mathbf{e}_l$, then $\mathbf{L} = \mathbf{E}_{ij}$, $\mathbf{M} = \mathbf{E}_{kl}$, and $\langle \mathbf{E}_{ij}, \mathbf{E}_{kl} \rangle = \delta_{ik}\delta_{jl} = 1$ if $i = k$, $j = l$, $= 0$ otherwise. Thus the induced basis **E** in L is orthonormal.

**4.3.** (a) $\delta(\mathbf{Q}) = \|\mathbf{A} - \mathbf{Q}\|^2 = \text{Tr}\ \{(\mathbf{A} - \mathbf{Q})(\mathbf{A}^T - \mathbf{Q}^T)\} = \text{Tr}\ (\mathbf{A}\mathbf{A}^T) - 2\,\text{Tr}\,(\mathbf{Q}\mathbf{A}^T) + \text{Tr}\,(\mathbf{Q}\mathbf{Q}^T) = \|\mathbf{A}\|^2 + \|\mathbf{1}\|^2 - 2\,\text{Tr}\,(\mathbf{Q}\mathbf{A}^T)$. (b) Part (a), together with $\mathbf{A} = \mathbf{V}\mathbf{R}$, immediately gives $\delta(\mathbf{Q}) = \|\mathbf{A}\|^2 + \|\mathbf{1}\|^2 - 2\,\text{Tr}\,(\mathbf{Q}\mathbf{R}^T\mathbf{V})$ (*). Since the tensor $\mathbf{P} = \mathbf{Q}\mathbf{R}^T$, being the product of two orthogonal tensors, is also orthogonal, one has $\min_{\mathbf{Q}\in\mathcal{O}} \delta(\mathbf{Q}) = \|\mathbf{A}\|^2 + \|\mathbf{1}\|^2 - 2\max_{\mathbf{P}\in\mathcal{O}} \Delta(\mathbf{P})$. (c) Clearly $\text{Tr}\,(\mathbf{P}\mathbf{V}) = \sum p_{ii}\lambda_i$. By Problem 3.20, $|p_{ii}| \le 1$ (no sum) for each i; since all the $\lambda_i$s are positive by the positive definiteness of **V**, one has $\Delta(\mathbf{P}) = \text{Tr}\,(\mathbf{P}\mathbf{V}) \le \sum \lambda_i = \text{Tr}\,(\mathbf{V}) = \Delta(\mathbf{1})$. (d) Referring to (*) above, one sees that the minimum of $\delta(\mathbf{Q})$ is obtained by making $\mathbf{Q}\mathbf{R}^T = \mathbf{1}$, i.e., by choosing $\mathbf{Q} = \mathbf{R}$. Thus the "minimum distance from **A** to **Q**" occurs when $\mathbf{Q} = \mathbf{R}$, and its value is $\delta(\mathbf{R}) = \|\mathbf{1}\|^2 + \|\mathbf{A}\|^2 - 2\text{Tr}\,\{(\mathbf{A}\mathbf{A}^T)^{1/2}\}$. Notice that if **A** happens to be orthogonal, then $\mathbf{V} = \mathbf{1}$, $\mathbf{A} = \mathbf{R}$, $\|\mathbf{A}\| = \|\mathbf{1}\|$, and $\min \delta(\mathbf{Q}) = \delta(\mathbf{R}) = 0$.

**4.4.** If **A** and **B** commute, then they have a common principal basis, which is also a principal basis for the symmetric tensor $\mathbf{C} = \mathbf{AB}$, which is therefore also positive definite, as is $\mathbf{C}^2$. Then $\mathrm{Tr}\{(\mathbf{BA}^2\mathbf{B})^{1/2}\} = \mathrm{Tr}\,\{(\mathbf{C}^2)^{1/2}\} = \mathrm{Tr}\,\mathbf{C} = \mathrm{Tr}\,\mathbf{AB} = \langle\mathbf{A}, \mathbf{B}\rangle$, so it follows that $\min \delta(\mathbf{P}, \mathbf{Q}) = \|\mathbf{A} - \mathbf{B}\|^2$. Note that this minimum vanishes when $\mathbf{A} = \mathbf{B}$, as indeed it should.

**4.5.** Let $c_{ijkl}$ be the components in an orthonormal basis of a 4-tensor $\underline{\underline{\mathbf{C}}}$ that is symmetric and has the first minor symmetry. By the major symmetry, $c_{ijkl} = c_{klij}$, so by the first minor symmetry, $c_{ijkl} = c_{lkij}$. Using the major symmetry again gives $c_{ijkl} = c_{ijlk}$, so $\underline{\underline{\mathbf{C}}}$ has the second minor symmetry. Similarly, if $\underline{\underline{\mathbf{C}}}$ has the major symmetry and the *second* minor symmetry, it also has the first minor symmetry. Can you find an argument that does not make use of components?

**4.6.** For any 2-tensor $\mathbf{A}$, $\|\underline{\underline{\mathbf{P}}}\,\mathbf{A}\|^2 = \|\mathbf{QAR}\|^2 = \mathrm{Tr}\,\,(\mathbf{QARR}^T\mathbf{A}^T\mathbf{Q}^T) = \mathrm{Tr}\,(\mathbf{QAA}^T\mathbf{Q}^T)$, so that by the invariance of the trace, $\|\underline{\underline{\mathbf{P}}}\,\mathbf{A}\|^2 = \|\mathbf{A}\|^2$, showing that $\underline{\underline{\mathbf{P}}}$ is an orthogonal 4-tensor.

**4.7.** $\underline{\underline{\mathbf{L}}}$ has the major symmetry, but neither minor symmetry. This is easy to show by using components.

**4.8.** If $\underline{\underline{\mathbf{A}}} = \mathbf{B}\otimes\mathbf{C}$, the components of the 4-tensor $\underline{\underline{\mathbf{A}}}$ in an orthonormal basis are $a_{ijkl} = b_{ij}c_{kl}$. In particular, the components of $\mathbf{1}\otimes\mathbf{1}$ are $\delta_{ij}\delta_{kl}$.

**4.9.** $g_{ijkl} = c_{ijpq}d_{pqkl}$.

**4.10.** In a given orthonormal basis, one may arbitrarily assign the values of twenty-one among the eighty-four components of a 4-tensor that has the major symmetry and both minor symmetries.

**4.11.** Let $\mathbf{e}_1, \ldots, \mathbf{e}_n$ be an orthonormal basis for $\mathsf{R}$, and let $\mathbf{E}_{ij} = \mathbf{e}_i\otimes\mathbf{e}_j$ be the tensors of the induced orthonormal basis for the associated space $\mathsf{L}$. Let $\underline{\underline{\mathbf{T}}}\mathbf{E}_{kl} = t_{ijkl}\,\mathbf{E}_{ij}$, where $t_{ijkl}$ are the components of the transformation 4-tensor $\underline{\underline{\mathbf{T}}}$ in in the basis $\mathbf{e}$. Then $t_{ijkl} = \langle\underline{\underline{\mathbf{T}}}\mathbf{E}_{kl}, \mathbf{E}_{ij}\rangle = \langle\mathbf{E}_{lk}, \mathbf{E}_{ij}\rangle = \delta_{il}\delta_{jk}$, independently of the chosen basis $\mathbf{e}$. Thus the components of $\underline{\underline{\mathbf{T}}}$ are indeed independent of the choice of orthonormal basis, and $\underline{\underline{\mathbf{T}}}$ is therefore an isotropic 4-tensor. Alternatively, if one sets $c^{\mathbf{e}}_{pqrs} = \delta_{ps}\delta_{qr}$ in the change-of-basis formula (4.25), one finds that $c^{\mathbf{f}}_{ijkl} = r^{\mathbf{e}}_{pi}r^{\mathbf{e}}_{qj}r^{\mathbf{e}}_{rk}r^{\mathbf{e}}_{sl}\,\delta_{ps}\delta_{qr} = r^{\mathbf{e}}_{pi}r^{\mathbf{e}}_{pl}r^{\mathbf{e}}_{qj}r^{\mathbf{e}}_{qk} = \delta_{il}\delta_{jk}$, so that a 4-tensor whose components are $\delta_{il}\delta_{jk}$ in one orthonormal basis $\mathbf{e}$ has these same components in all orthonormal bases.

**4.12.** This follows immediately from the change-of-basis formula (4.25) for the components of a 4-tensor.

**4.13.** $\underline{\underline{\mathbf{T}}}\mathbf{A} = \lambda\,\mathbf{A}$ is equivalent to $\mathbf{A}^T - \lambda\,\mathbf{A} = \mathbf{O}$. In the 2-dimensional space $\mathsf{R}_2$, this in turn is equivalent to the four scalar equations $(1 - \lambda)a_{11} = 0$, $a_{21} - \lambda a_{12} = 0$, $a_{12} - \lambda a_{21} = 0$, $(1 - \lambda)\,a_{22} = 0$. If $\lambda \neq 1$, then $a_{11} =$

$a_{22} = 0$. Since $(1 - \lambda^2)a_{21} = 0$, $\lambda \neq \pm 1$ implies $a_{21} = a_{12} = 0$ as well. Thus if $\lambda \neq \pm 1$, then $\mathbf{A} = 0$, so $\lambda = \pm 1$ are the only possible eigenvalues of $\underline{\underline{\mathbf{T}}}$. Suppose $\lambda = 1$. Then $a_{11}$ and $a_{22}$ are arbitrary, and $a_{21} = a_{12}$, with $a_{12}$ arbitrary. Thus any non-null symmetric $\mathbf{A}$ is an eigentensor of $\underline{\underline{\mathbf{T}}}$ corresponding to the eigenvalue $\lambda = 1$. The space $\mathsf{L}$ of all 2-tensors on $\mathsf{R}_2$ has dimension four; there are three linearly independent symmetric $\mathbf{A}$s, so the dimension of the invariant linear manifold corresponding to $\lambda = 1$ is three. If $\lambda = -1$, then necessarily $a_{11} = a_{22} = 0$, and $a_{21} = -a_{12}$, so $\mathbf{A}$ is skew-symmetric, and the corresponding invariant linear manifold has dimension one. The two invariant linear manifolds are orthogonal. Can you generalize this result to an arbitrary finite-dimensional real Euclidean space?

**4.14.** Suppose there is a non-null 2-tensor $\mathbf{X}$ such that $(\mathbf{A}\otimes\mathbf{A})\mathbf{X} = \lambda\, \mathbf{X}$ for some scalar $\lambda$. Then $\langle \mathbf{A}, \mathbf{X}\rangle\, \mathbf{A} = \lambda\, \mathbf{X}$. If $\mathbf{X}$ is orthogonal to $\mathbf{A}$, then $\lambda = 0$, and conversely; thus $\lambda = 0$ is an eigenvalue of $\mathbf{A}\otimes\mathbf{A}$, and any 2-tensor $\mathbf{X}$ orthogonal to $\mathbf{A}$ is a corresponding eigentensor. If $\lambda \neq 0$, then $\mathbf{X}$ is a scalar multiple of $\mathbf{A}$, and necessarily $\lambda = \langle \mathbf{A}, \mathbf{A}\rangle = \|\mathbf{A}\|^2$. Compare with Problem 3.5. Spectral representation: $\mathbf{A}\otimes\mathbf{A} = \|\mathbf{A}\|\ \mathbf{E}$, where $\mathbf{E} = (1/\|\mathbf{A}\|)\ \mathbf{A}$. (Really!)

**4.15.** Suppose $\underline{\underline{\mathbf{C}}}\,\mathbf{A} = \lambda\mathbf{A}$ for some non-null 2-tensor $\mathbf{A}$ and some scalar $\lambda$, and let $\underline{\underline{\mathbf{D}}}$ be the 4-tensor defined by $\underline{\underline{\mathbf{D}}} = \mathbf{1}\otimes\mathbf{1}$. Then $\underline{\underline{\mathbf{D}}}\mathbf{A} = \mu\mathbf{A}$, where $\mu = (\lambda - \alpha)/\beta$, so $\mathbf{A}$ is an eigentensor of $\underline{\underline{\mathbf{D}}}$ corresponding to the eigenvalue $\mu$. By the results in Problem 4.14, the eigenvalues of $\underline{\underline{\mathbf{D}}}$ are $\lambda = 0$ and $\lambda = \|\mathbf{1}\|^2 = \mathrm{n}$, where n is the dimension of the underlying vector space. The set of eigentensors corresponding to $\mu = 0$ consists of all 2-tensors orthogonal to $\mathbf{1}$, which is the set of all 2-tensors with zero trace. The eigentensors associated with $\mu = \mathrm{n}$ are the scalar multiples of $\mathbf{1}$. The values of $\lambda$ that correspond to $\mu = 0$ and $\mu = \mathrm{n}$ are $\lambda = \alpha$ and $\lambda = \alpha + \beta\mathrm{n}$.

**4.16.** In terms of components in an orthonormal basis, $\underline{\underline{\mathbf{C}}}\,\mathbf{A} = \mathbf{O}$ is equivalent to $c_{ijkl}a_{kl} = 0$. Since $\underline{\underline{\mathbf{C}}}$ has the second minor symmetry, one also has $c_{ijkl}(a_{kl} + a_{lk}) = 0$. Choose $\mathbf{A}$ to be a non-null skew-symmetric 2-tensor, so that $a_{kl} + a_{lk} = 0$. Then $\underline{\underline{\mathbf{C}}}\,\mathbf{A} = \mathbf{O}$, so that $\underline{\underline{\mathbf{C}}}$ is singular.

**4.17.** (a) $\langle \underline{\underline{\mathbf{C}}}\mathbf{A}, \mathbf{A}\rangle = c_{ijkl}a_{ij}a_{kl}$. If $\underline{\underline{\mathbf{C}}} = \underline{\underline{\mathbf{1}}}$, then $\langle \underline{\underline{\mathbf{C}}}\mathbf{A}, \mathbf{A}\rangle = \|\mathbf{A}\|^2$. If $\underline{\underline{\mathbf{C}}} = \mathbf{1}\otimes\mathbf{1}$, then $\langle \underline{\underline{\mathbf{C}}}\mathbf{A}, \mathbf{A}\rangle = a_{ii}a_{kk} = \{\mathrm{Tr}\,(\mathbf{A})\}^2 = \langle \mathbf{A}, \mathbf{1}\rangle^2$; $\underline{\underline{\mathbf{1}}}$ is positive definite, but $\mathbf{1}\otimes\mathbf{1}$ is not, since $\langle \mathbf{A}, \mathbf{1}\rangle$ vanishes for any traceless $\mathbf{A}$. (b) $c_{ijkl}u_iu_kv_jv_l > 0$. $\underline{\underline{\mathbf{1}}}$ is strongly elliptic, but $\mathbf{1}\otimes\mathbf{1}$ is not, since $\mathrm{Tr}\,(\mathbf{u}\otimes\mathbf{v}) = 0$ if $(\mathbf{u}, \mathbf{v}) = 0$. Clearly positive definiteness implies strong ellipticity. Can you find an example to show that the converse is false?

## Chapter 5

**5.1.** A line segment in D may be represented parametrically by the equation $\mathbf{x} = \mathbf{x}' + \mathbf{m}\alpha$, $0 \leq \alpha \leq \alpha_1$, where $\alpha$ is a "tracing parameter" for the line, and $\mathbf{m}$ is a given unit vector that specifies the direction of the line. Under the homogeneous deformation (5.2), the image of the line segment is given by $\mathbf{y} = \mathbf{F}\mathbf{x} = \mathbf{y}' + \mathbf{n}\alpha$, $0 \leq \alpha \leq \alpha_1$, where $\mathbf{y}' = \mathbf{F}\mathbf{x}'$ and the new direction vector $\mathbf{n}$ is given by $\mathbf{n} = \mathbf{F}\mathbf{m}$. It follows that $\mathbf{n} \neq \mathbf{o}$ since $\mathbf{F}$ is nonsingular. A similar argument using two tracing parameters can be used to show that the deformation (5.2) carries a plane into a plane.

**5.2.** Using the notation of (5.5), one may calculate the change in the square of the length of the segment $\mathbf{q}$, regarded as given in D$'$, as it is carried to D by the deformation inverse to (5.2) as follows: $|\mathbf{p}|^2 = |\mathbf{F}^{-1}\mathbf{q}|^2 = (\mathbf{q},\ \mathbf{F}^{-T}\mathbf{F}^{-1}\mathbf{q}) = (\mathbf{q},\ (\mathbf{F}\mathbf{F}^{T})^{-1}\mathbf{q}) = (\mathbf{q},\ \mathbf{V}^{-2}\mathbf{q})$, so that $|\mathbf{q}|^2 - |\mathbf{p}|^2 = 2(\mathbf{q},\ \mathbf{H}\mathbf{q})$, where $\mathbf{H} = (1/2)\ (\mathbf{1} - \mathbf{V}^{-2}) = (1/2)\ (\mathbf{1} - (\mathbf{F}\mathbf{F}^{T})^{-1})$. The symmetric tensor $\mathbf{H}$ is called the *Eulerian* strain tensor. The associated relative elongation is $\Delta(\mathbf{q}) = (|\mathbf{p}|/|\mathbf{q}|) - 1 = \{1 - 2\,(\mathbf{q}, \mathbf{H}\mathbf{q})/(\mathbf{q}, \mathbf{q})\}^{1/2} - 1$.

**5.3.** Clearly the maximum elongation of $\mathbf{p}$ occurs when $\mathbf{p}$ makes $(\mathbf{p}, \mathbf{E}\mathbf{p})/(\mathbf{p}, \mathbf{p})$ a maximum. By the argument used to establish Rayleigh's principle, this ratio is greatest when $\mathbf{p}$ is a multiple of the eigenvector $\mathbf{e}_3$ associated with the largest eigenvalue, say $\epsilon_3$, of the symmetric Lagrangian strain tensor $\mathbf{E}$. Similarly, the minimum relative elongation occurs if $\mathbf{p} = \alpha\mathbf{e}_1$, where $\mathbf{e}_1$ is the eigenvector of $\mathbf{E}$ corresponding to the smallest eigenvalue, say $\epsilon_1$.

**5.4.** Note first that $\mathbf{A}^T$ is isotropic if $\mathbf{A}$ is isotropic. Since $\mathbf{S} = (\mathbf{A} + \mathbf{A}^T)/2$ and $\Omega = (\mathbf{A} - \mathbf{A}^T)/2$, it follows immediately that $\mathbf{Q}\mathbf{S}\mathbf{Q}^T = \mathbf{S}$, $\mathbf{Q}\Omega\mathbf{Q}^T = \Omega$ for all proper orthogonal $\mathbf{Q}$. Thus $\mathbf{S}$ and $\Omega$ are isotropic 2-tensors. (b) Let $a_{ij}$ be the components of $\Omega$ in an orthonormal basis $\mathbf{e}$, and choose $\mathbf{Q}$ so that its matrix of components in $\mathbf{e}$ is that appearing on the right side of (5.24). For this $\mathbf{Q}$, the relation $\mathbf{Q}\Omega\mathbf{Q}^T = \Omega$ implies that $a_{12} = a_{23} = 0$. Similar choices of $\mathbf{Q}$ lead to $a_{ij} = 0$ for all other pairs i, j with $i \neq j$. This, together with the skew-symmetry of $\Omega$, implies that $\Omega = \mathbf{O}$. (c) Since $\mathbf{S}$ is a scalar multiple of the identity by Proposition 5.1, it follows that the same is true of $\mathbf{A}$.

**5.5.** (a) Let $\Omega = \begin{pmatrix} 0 & \beta \\ -\beta & 0 \end{pmatrix}$, and choose $\mathbf{Q} = \begin{pmatrix} \cos\varphi & -\sin\varphi \\ \sin\varphi & \cos\varphi \end{pmatrix}$. Direct calculation shows that $\mathbf{Q}\Omega\mathbf{Q}^T = \Omega$, so that any skew-symmetric tensor on $R_2$ is isotropic. (b) The most general isotropic tensor on $R_2$ is $\begin{pmatrix} \alpha & \beta \\ -\beta & \alpha \end{pmatrix}$, where

$\alpha$ and $\beta$ are arbitrary scalars. (c) With $\Omega$ as in part (a), choose $\mathbf{Q} = \begin{pmatrix} 0 & 1 \\ 1 & 0 \end{pmatrix}$, which is orthogonal, but not proper. Direct calculation shows that $\mathbf{Q}\Omega\mathbf{Q}^{\mathrm{T}} = -\Omega$, so $\Omega$ is isotropic if and only if it vanishes.

**5.6.** If the components of $\mathbf{x}$ are the same in all orthonormal bases, the change-of-basis formula (1.9) gives $(\rho_{jk} - \delta_{jk})\xi_k = 0$, where $\rho_{jk}$ are the components of an arbitrary proper orthogonal tensor $\mathbf{R}$ in an orthonormal basis, and the $\xi_k$s are the components of $\mathbf{x}$ in this basis. By suitable choices of $\mathbf{R}$, one can show successively that $\xi_1 = \xi_2 = \cdots = \xi_n = 0$.

**5.7.** Let $\mathbf{f}$ be the natural basis, so that $\underline{A} = \underline{A}^{\mathbf{f}} = \begin{pmatrix} 1 & 8/3 \\ 0 & 1 \end{pmatrix}$. Proceeding as in Problem 3.23, one finds that $(\underline{A}^{\mathbf{f}})^{\mathrm{T}}\underline{A}^{\mathbf{f}} = \begin{pmatrix} 1 & 8/3 \\ 8/3 & 73/9 \end{pmatrix}$, $\underline{U}^{\mathbf{f}} = \begin{pmatrix} 3/5 & 4/5 \\ 4/5 & 41/15 \end{pmatrix}$. Then $\mathbf{A} = \mathbf{RU}$, where $\underline{R}^{\mathbf{f}} = \begin{pmatrix} 3/5 & 4/5 \\ -4/5 & 3/5 \end{pmatrix}$. One also finds that $\mathbf{A} = \mathbf{VR}$, where $\underline{V}^{\mathbf{f}} = \begin{pmatrix} 41/15 & 4/5 \\ 4/5 & 3/5 \end{pmatrix}$

**5.8.** The 4-tensor $\mathbf{1}\otimes\mathbf{1}$ has components $c_{ijkl} = \delta_{ij}\delta_{kl}$. Thus $c_{ijkl} - q_{pi}q_{qj}q_{rk}q_{sl}c_{pqrs} = \delta_{ij}\delta_{kl} - q_{pi}q_{pj}q_{rk}q_{rl} = \delta_{ij}\delta_{kl} - \delta_{ij}\delta_{kl} = 0$, which is the component version of (5.26). The argument is similar for the 4-tensor $\underline{1}$, whose components are $c_{ijkl} = \delta_{ik}\delta_{jl}$.

**5.9.** According to Problem 4.17, one must compute $\langle \mathbf{A}, \underline{\underline{\mathbf{C}}}\,\mathbf{A}\rangle$ for an arbitrary symmetric 2-tensor $\mathbf{A}$. With $\underline{\underline{\mathbf{C}}}$ given by (5.27), one has $\langle \mathbf{A}, \underline{\underline{\mathbf{C}}}\,\mathbf{A}\rangle = 2\mu \,\|\mathbf{A}\|^2 + \lambda\,\{\mathrm{Tr}\,(\mathbf{A})\}^2$. Suppose $\langle \mathbf{A}, \underline{\underline{\mathbf{C}}}\,\mathbf{A}\rangle$ is positive for all symmetric non-null $\mathbf{A}$. By first choosing a traceless $\mathbf{A}$, one sees that $\mu > 0$ is necessary. By next choosing $\mathbf{A}$ so that, in an orthonormal basis $\mathbf{e}$, one has $a^{\mathbf{e}}_{11} = 1$, while all other components of $\mathbf{A}$ vanish, one finds that $2\mu + \lambda > 0$ is also necessary. Conversely, one can show easily that the two conditions $\mu > 0$, $2\mu + \lambda > 0$ are also *sufficient* for the positivity of $\langle \mathbf{A}, \underline{\underline{\mathbf{C}}}\,\mathbf{A}\rangle$.

**5.10.** The results asserted in parts (a) and (b) come from the chain rule for partial differentiation.

**5.11.** First, set $\varphi(\mathbf{A}) = \mathrm{I}_1(\mathbf{A})$. Then in a basis $\mathbf{e}$, one has $\varphi^{\mathbf{e}}(a^{\mathbf{e}}_{11}, a^{\mathbf{e}}_{12}, a^{\mathbf{e}}_{21}, a^{\mathbf{e}}_{22}) = a^{\mathbf{e}}_{11} + a^{\mathbf{e}}_{22}$, whence $\partial\varphi^{\mathbf{e}}/\partial a^{\mathbf{e}}_{ij} = \delta_{ij}$, so $\varphi_{\mathbf{A}}(\mathbf{A}) = \mathbf{1}$. Next, set $\varphi(\mathbf{A}) = \mathrm{I}_2(\mathbf{A}) = \det \mathbf{A}$, so that $\varphi^{\mathbf{e}}(a^{\mathbf{e}}_{11}, \ldots, a^{\mathbf{e}}_{22}) = a^{\mathbf{e}}_{11}a^{\mathbf{e}}_{22} - a^{\mathbf{e}}_{12}a^{\mathbf{e}}_{21}$. Then $\partial\varphi^{\mathbf{e}}/\partial a^{\mathbf{e}}_{11} = a^{\mathbf{e}}_{22}$, $\partial\varphi^{\mathbf{e}}/\partial a^{\mathbf{e}}_{22} = a^{\mathbf{e}}_{11}$, $\partial\varphi^{\mathbf{e}}/\partial a^{\mathbf{e}}_{12} = -a^{\mathbf{e}}_{21}$, $\partial\varphi^{\mathbf{e}}/\partial a^{\mathbf{e}}_{21} = -a^{\mathbf{e}}_{12}$. Let $\mathbf{B} = (\mathbf{1}\otimes\mathbf{1})\mathbf{A} - \mathbf{A}^{\mathrm{T}}$. It is easily verified that $b^{\mathbf{e}}_{11} = a^{\mathbf{e}}_{22}$, $b^{\mathbf{e}}_{12} = -a^{\mathbf{e}}_{21}$, $b^{\mathbf{e}}_{21} = -a^{\mathbf{e}}_{12}$, $b^{\mathbf{e}}_{22} = a^{\mathbf{e}}_{11}$. It follows that $\varphi_{\mathbf{A}}(\mathbf{A}) = \mathbf{B}$.

**5.12.** Referring to (5.55), we write $\varphi(\mathbf{A}) = \psi(\mathrm{I}_1(\mathbf{A}), \mathrm{I}_2(\mathbf{A})$. By the chain rule, $\varphi_{\mathbf{A}} = \partial\psi/\partial\mathrm{I}_1\partial\mathrm{I}_1/\partial\mathbf{A} + \partial\psi/\partial\mathrm{I}_2\partial\mathrm{I}_2/\partial\mathbf{A}$. Using the result of Problem 5.11 then gives $\varphi_{\mathbf{A}} = \partial\psi/\partial\mathrm{I}_1\mathbf{1} + \partial\psi/\partial\mathrm{I}_2\,[(\mathbf{1}\otimes\mathbf{1})(\mathbf{A} - \mathbf{A}^{\mathrm{T}}]$.

**5.13.** $\varphi(\mathbf{QAQ}^{\mathrm{T}}) = \langle \mathbf{QAQ}^{\mathrm{T}}, \underline{\underline{\mathbf{C}}}(\mathbf{QAQ}^{\mathrm{T}})\rangle = \langle \underline{\underline{\mathbf{Q}}}\,\mathbf{A}, \underline{\underline{\mathbf{C}}}\,\underline{\underline{\mathbf{Q}}}\,\mathbf{A}\rangle$, where $\underline{\underline{\mathbf{Q}}}$ is the 4-rotation associated with the orthogonal 2-tensor $\mathbf{Q}$. By the isotropy of the 4-tensor $\underline{\underline{\mathbf{C}}}$, one has $\underline{\underline{\mathbf{C}}}\,\underline{\underline{\mathbf{Q}}} = \underline{\underline{\mathbf{Q}}}\,\underline{\underline{\mathbf{C}}}$, so that $\varphi(\mathbf{QAQ}^{\mathrm{T}}) = \langle \underline{\underline{\mathbf{Q}}}\mathbf{A}, \underline{\underline{\mathbf{Q}}}(\underline{\underline{\mathbf{C}}}\mathbf{A})\rangle = \langle \mathbf{A}, \underline{\underline{\mathbf{Q}}}^{\mathrm{T}}\underline{\underline{\mathbf{Q}}}(\underline{\underline{\mathbf{C}}}\,\mathbf{A})\rangle = \langle \mathbf{A}, \underline{\underline{\mathbf{C}}}\,\mathbf{A}\rangle = \varphi(\mathbf{A})$. Thus $\varphi$ is an isotropic scalar-valued function.

**5.14.** Since $\underline{\mathrm{A}}$ and $\underline{\mathrm{B}}$ commute, the system has classical normal modes.

# INDEX

Acoustic tensor, 76 (Problem 4.18)
Additive decomposition of a tensor, 57
Angle
between arrows, 9
between deformed segments in a continuum, 81
Arrows, 1

Basis
change of, 7, 30, 73
definition, 5
induced, 68
natural, 6
orthonormal, 10
principal, 46
Bessel inequality. *See* Inequality

Cayley-Hamilton theorem, 40–41 (Problem 2.18)
Change-of-basis formula
for components of a four-tensor, 73
for components of a linear transformation, 30–31, 56–57
for components of a vector, 7
for matrix of components of a linear transformation, 30–31
for representation of a scalar function of tensors, 90–91
Characteristic polynomial. *See also* Polynomials
complex zeros of, 32–35 (Propositions 2.1, 2.2), 39–40 (Problem 2.13), 52–53
real zeros of, 32 (Proposition 2.2), 52
Complex Euclidean space, 14
Components
of a linear transformation, 27
of a vector, 5
Continuous functions
as a Euclidean space, 12
space C of, 3
Continuous medium, 78
Contraction transformation, 63 (Problem 3.16)
Cramer's Rule, 22, 100
Crystals
martensitic transformations in, 75 (Problem 4.4)
material symmetries of, 90
phase transitions in, 76 (Problem 4.17)

Decompositions of tensors
additive, 57
polar, 57–59, 64–65 (Problem 3.28)
Definite. *See* Positive definite; Symmetric tensors
Deformations
definition, 78
homogeneous, 79–80
infinitesimal, 82
rigid, 82
Determinant
definition, 99–100
invariance of, 29, 31
of a linear transformation, 29
Dimension, 5
Displacement
gradient, 82
vector, 82
Dynamical system
approximate approach to, 93–94, 97–98 (Problem 5.15)
classical normal modes in, 93
convenience hypothesis for, 93
degrees of freedom of, 93

Eigentensors, 70, 76 (Problems 4.13–4.15), 87
Eigenvalues, 24
Eigenvectors, 24
Elasticity tensor
definition, 69, 84
isotropic, 86–90
Euclidean space. *See* Real Euclidean space; Complex Euclidean space

Four-tensors
change-of-basis formula for components of, 73
components of, 69
definition, 68–69, 73
elasticity tensor, 69, 76 (Problem 4.18)
four-rotations, 71
four-tensor product of two-tensors, 69
identity four-tensor, 69
isotropic, 86–90
major symmetry of, 71
minor symmetries of, 71
orthogonal, 71
positive definite, 76 (Problem 4.17)

strongly elliptic, 76 (Problem 4.17)
symmetric, 70
transpose of, 70
transposition four-tensor, 69
Fourier series, 13
Fredholm
alternative for arbitrary tensors, 65 (Problem 3.29)
alternative for symmetric tensors, 61 (Problem 3.13)
integral equation 19–20, 26
Free index, 7

Gram-Schmidt process, 11, 16 (Problem 1.17)
Group of tensors, 90

Heat conduction
Fourier's law for, 84
heat flux in, 83
isotropic, 84
temperature gradient in, 83

Image
of a transformation, 20
of a vector under a transformation, 18
Inequality
Bessel, 61 (Problem 3.12)
Schwarz, 10
triangle, 10
Integral equation 19–20, 26, 38 (Problem 2.8)
Invariant. *See* Invariant linear manifold; Scalar invariant
Invariant linear manifold. *See also* Linear manifold
definition, 23
irreducible, 27, 40 (Problem 2.14)
one-dimensional, 24
two-dimensional, 26, 53
Inverse
components of, 29
definition, 23
linearity of, 23
Isotropic
four-tensor, 86–90, 96 (Problems 5.8, 5.9)
scalar-valued function of a tensor, 91
two-tensor, 84–86, 95 (Problem 5.4)
Isotropy, 84–90

Jordan canonical form. *See* Matrices

Kernel. *See* Null space
Kronecker delta, 11

Legendre polynomials. *See* Polynomials
Linear dependence, 4
Linear function, 43, 59–60 (Problem 3.1)
Linear independence, 4
Linear manifold, 8. *See also* Invariant linear manifold
Linear transformations. *See* Transformations
Linear vector space
complex, 2
of complex columns ($\mathbf{C_n}$), 3
of continuous functions ($\mathbf{C}$), 3
definition, 1–2
Euclidean, 9, 14
finite dimensional, 5
infinite dimensional, 5
of linear transformations ($\mathbf{L}$), 20, 67
real, 2
of real columns ($\mathbf{R_n}$), 2
of real polynomials ($\mathbf{P_n}$), 4
role of scalars in, 1
of square-summable sequences ($\mathbf{l_2}$), 17 (Problem 1.20)
unitary, 14
vectors in, 1

Material symmetry, 90
Matrices
arithmetic of, 99
of components of linear transformations, 27
determinant of, 29, 99–100
diagonal, role of, 35 (Proposition 2.3)
Jordan canonical form of, 36
rule for multiplying, 28, 99
Minimum principle, 62 (Problem 3.14)
Multiplicity
of eigenvalues, algebraic, 35, 47 (Proposition 3.3)
of eigenvalues, geometric, 35, 47 (Proposition 3.3)
of eigenvalues of an orthogonal tensor, 63 (Problem 3.21)

Negative definite. *See* Symmetric tensors
Null space, 20

Orthogonal
group, 90
pairs of vectors, 10
tensors, 51
transformations, 51
Orthogonal tensors
definition, 51
proper, improper, 52
reflection by, 53–55
representation of, 55 (Proposition 3.5)
rotation by, 55–56, 72
Orthonormal basis. *See* Basis

Polar decomposition of a tensor
left, right 58
for nonsingular tensors, 58 (Proposition 3.6)
for singular tensors, 59, 64 (Problem 3.28)
Polynomials
characteristic, 32–35, 39–40 (Problem 2.13), 40–41 (Problem 2.18)

Legendre, 16 (Problem 1.18)
in linear transformations, 40–41 (Problem 2.18)
trigonometric, 12–13
as a vector space, 4
Positive definite. *See* Symmetric tensors; Quadratic forms
Principal basis
for commuting symmetric tensors, 47–48 (Proposition 3.4), 93–94
definition, 46
Product
dot, 9
scalar, 9, 67–68
tensor product of vectors, 44
of two linear transformations, 20
Projection, 61 (Problem 3.12)

Quadratic form
definition, 48
positive definite, 61 (Problem 3.9), 66 (Problem 3.30)

Range, 20
Range convention, 7
Rank. *See* Transformations
Rayleigh's principle, 62 (Problems 3.14, 3.15)
Real Euclidean space, 9
Relative elongation, 80
Rotation
four-rotation, 71
of two-tensors, 72

Scalar invariants
definition, 91
examples, 31, 91
fundamental, 92
representation of, 91–92
Scalar product. *See* Product
Scalars. *See* Linear vector spaces
Schwarz inequality. *See* Inequality
Shear, 96 (Problem 5.7)
Skew-symmetric tensors, 45, 57, 61 (Problem 3.10)
Span, 8
Spectral formula, 46–47
Strain
components of, 82
definitions, 80, 95 (Problem 5.2)
Eulerian and Lagrangian, 95 (Problem 5.2)
infinitesimal, 82
principal, 80, 95 (Problem 5.3)
Stress in a solid, 69, 83–84
Stretch tensors
left, right in polar decomposition, 59
for a continuum, 79, 82–83
Subspace, 8
Summation convention, 7
Symmetric tensors
definition, 45
eigenvalues and eigenvectors of, 45–46
geometric effect of, 49
negative definite, 48
positive definite, 48
representation of, 46, 47
square roots of, 49, 58, 64 (Problem 3.27)
zeros of characteristic polynomials of, 45 (Proposition 3.1)

Tensors. *See also* Four-tensors; Two-tensors; Transformations
acoustic, 76 (Problem 4.18)
Cartesian, 42, 73
elasticity, 69, 84
induced basis for, 68
isotropic, 84–90
length of, 68
orthogonal, 51
representation of, 27, 67
scalar product of, 67–68
scalar-valued functions of, 90
skew-symmetric, 45, 57, 61 (Problem 3.10)
strain, 80, 95 (Problem 5.2)
stretch, 79
symmetric, 45–51
Trace, 31, 40 (Problem 2.15), 57
Transformations. *See also* Tensors
commutative, 20, 47–48 (Proposition 3.4), 93
components of, 27
contracting, 63 (Problem 3.16)
definition, 18
linear, 18
matrix of components of, 27
nonsingular, 21–23, 29–30, 37 (Problem 2.4)
rank of, 60 (Problem 3.6)
rank-one connection of, 60 (Problem 3.7)
representation of, 27, 67
singular, 22, 39 (Problem 2.10), 59, 64 (Problem 3.28)
skew-symmetric, 45, 57, 61 (Problem 3.10)
symmetric, 45–51
Transpose
of matrix, 100
of product of tensors, 43, 60 (Problem 3.2)
of tensor, 42–45
of tensor product of vectors, 44
Triangle inequality. *See* Inequality
Two-tensors, 69, 73

Unitary space, 14

Vector spaces. *See* Linear vector space
Vectors. *See* Linear vector space

Wave propagation, 76–77 (Problem 4.18)